Power Electronics

Lecture Notes of Power Electronics Course

By
Dr. Hidaia Mahamed Alassouli

Introduction

This book includes my lecture notes for power electronics course course. The characteristics and operation of electronic power devices, firing circuits, and driving circuits for power converters are described and implemented practically in the laboratory. Uncontrolled and controlled, single phase rectifiers are used in various electrical power applications. DC to DC power conversion circuits are investigated. Circuit simulation and practical laboratories are utilized to reinforce concepts.

The book is divided to different learning parts

- Part1- Describe the characteristics and operation of electronic power devices.
- Part2- Describe firing and driving circuits for power electronic converters.
- Part3- Analyse the use of uncontrolled and controlled single-phase rectifiers in various electrical power applications.
- Part4- Investigate the DC-to-DC power conversion circuits used in power applications.

Part1: Describe the characteristics and operation of electronic power devices.

1. Describe diode characteristics, types (power diode, general-purpose, and fast recovery), and connections (series, parallel and freewheeling).
2. Describe thyristor characteristics, two-transistor model, and purpose of di/dt and dv/dt protection.
3. Describe the power MOSFET and IGBT characteristics.
4. Compare electronic power devices in terms of various power converter applications, frequency of operation (switching speed), rating, and switching power losses.

Part 2: Describe firing and driving circuits for power electronic converters.

1. Describe ideal and non-ideal properties of operational amplifiers. Determine the operation of various related circuits (inverting and non-inverting amplifiers, buffer amplifier, summing amplifier)
2. Describe the use of an operational amplifier for PWM generation, for triangular and sine wave generation, as a comparator, and its integration into a 555 timer.
3. Explore other basic firing and driving circuits by focusing on requirements and control features such as based on specific power devices and operational amplifier.

Part 3: Analyse the use of uncontrolled and controlled single-phase rectifiers in various electrical power applications.

1. Determine the performance characteristics of uncontrolled single-phase, half-wave and full-wave rectifiers, with resistive and inductive loads.
2. Determine the performance characteristics of controlled single-phase, half-wave and full-wave rectifiers with resistive and inductive loads.
3. Determine the change in power factor when using uncontrolled and controlled rectifiers. Define input distortion and displacement factor.
4. Describe how power inversion may be achieved by varying the firing angle in controlled rectifiers.

Part 4: Investigate the DC-to-DC power conversion circuits used in power applications.

1. State the principle of step-down and step-up operations.
2. Explain the DC chopper classification and describe switch-mode regulators
3. Explain the operation of buck, boost
4. Explain the operation buck-boost regulators.

A. Part1: Describe the characteristics and operation of electronic power devices.

1. Describe diode characteristics, types (power diode, general-purpose, and fast recovery), and connections (series, parallel and freewheeling).
2. Describe thyristor characteristics, two-transistor model, and purpose of di/dt and dv/dt protection.
3. Describe the power MOSFET and IGBT characteristics.
4. Compare electronic power devices in terms of various power converter applications, frequency of operation (switching speed), rating, and switching power losses.

EGN 1103   Electrical / Electronics Department   ADMC

Electronics Devices

What are Semiconductors?

Semiconductors are materials which have a conductivity between **conductors** (generally metals) and nonconductors or **insulators** (such as most ceramics). Semiconductors can be pure elements, such as silicon or germanium, or compounds such as gallium arsenide or cadmium selenide. In a process called doping, small amounts of impurities are added to pure semiconductors causing large changes in the conductivity of the material.

Due to their role in the fabrication of electronic devices, semiconductors are an important part of our lives. Imagine life without electronic devices. There would be no radios, no TV's, no computers, no video games, and poor medical diagnostic equipment. Although many electronic devices could be made using vacuum tube technology, the developments in semiconductor technology during the past 50 years have made electronic devices smaller, faster, and more reliable. Think for a minute of all the encounters you have with electronic devices. How many of the following have you seen or used in the last twenty-four hours? Each has important components that have been manufactured with electronic materials.

microwave oven	electronic balance	video games
radio	television	VCR
watch	CD player	stereo
computer	lights	air conditioner
calculator	telephone	musical greeting cards
diagnostic equipment	clock	refrigerator
car	security devices	stove

Advances in the field of electronics can continue to improve our lives. Learning about electronic materials can help you understand and be able to participate in the fields of communication, computers, medicine, the basic sciences and engineering. All of these fields use electronics extensively.

1/12/2016   1   Nasser

EGN 1103   Electrical / Electronics Department   ADMC

The Doping of Semiconductors

The addition of a small percentage of foreign atoms in the regular crystal lattice of silicon or germanium produces dramatic changes in their electrical properties, producing n-type and p-type semiconductors.

Pentavalent impurities
Impurity atoms with 5 valence electrons produce n-type semiconductors by contributing extra electrons.

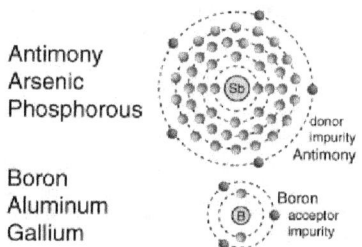

Antimony
Arsenic
Phosphorous

Boron
Aluminum
Gallium

Trivalent impurities
Impurity atoms with 3 valence electrons produce p-type semiconductors by producing a "hole" or electron deficiency.

P- and N- Type Semiconductors

Nasser

N-Type Semiconductor

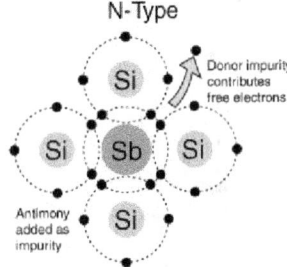

The addition of pentavalent impurities such as antimony, arsenic or phosphorous contributes free electrons, greatly increasing the conductivity of the intrinsic semiconductor. Phosphorous may be added by diffusion of phosphine gas (PH3).

P-Type Semiconductor

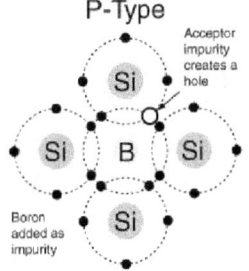

The addition of trivalent impurities such as boron, aluminum or gallium to an intrinsic semiconductor creates deficiencies of valence electrons, called "holes". It is typical to use B_2H_6 diborane gas to diffuse boron into the silicon material.

Diode

A semiconductor diode is a non-linear device whose most outstanding feature is the fact that, basically, current is only allowed to flow in one direction. The diode is built by joining together two semiconductor materials: an N-type material (rich in negative carriers or free electrons) and a P-type material (rich in positive carriers or holes). The area of contact is called the junction. For this reason the diode is commonly referred to as a **PN Junction**.

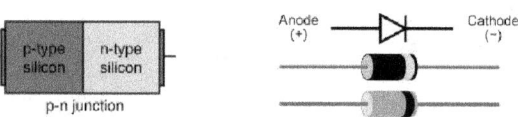

Current passing through a **diode** can only go in one direction, called the forward direction. Current trying to flow the reverse direction is blocked. They're like the one-way valve of electronics. If the voltage across a **diode** is negative, no current can flow*, and the ideal **diode** looks like an open circuit.

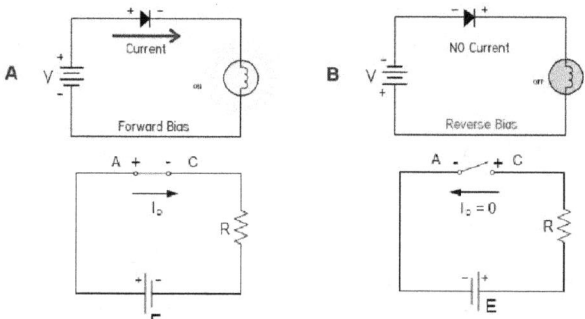

Forward bias

When the positive terminals of the supply are connected to the Anode and the minus is connected to the Cathode, the diode acts as a closed switch. The current flows through it.

Reverse bias

When the positive terminals of the supply are connected to the Cathode and the minus is connected to the Anode, the diode acts as an open switch. The current does not flows through it.

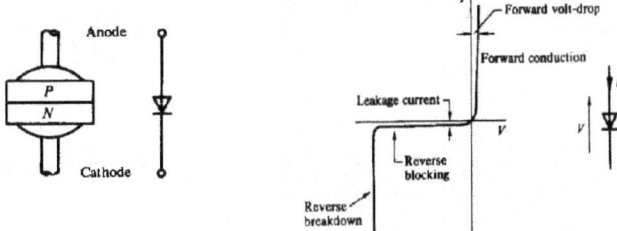

• **Peak Inverse voltage (PIV):** Is the maximum voltage that a diode can withstand only so much voltage before it breaks down. So if the PIV is exceeded than the PIV rated for the diode, then the diode will conduct in both forward and reverse bias and the diode will be immediately destroyed.

• **Maximum Average Current:** Is the average current that the diode can carry.

• **Reverse Recovery Time**

When the diode is switched from forward to reverse bias, the current does not immediately stop due to the excess carriers in the depletion region at the time of switching. The reverse recovery time is measured as the time delay between switching and when the current reaches ten percent of its maximum reverse value.

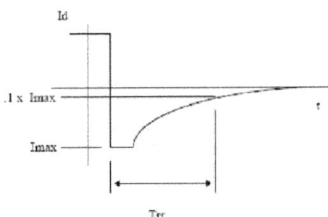

Graph of reverse recovery time of diode

| EGN 1103 | Electrical / Electronics Department | ADMC |

LED

A **light-emitting diode (LED)** is a two-lead semiconductor light source. It is a pn junction diode, which emits light when activated. When a suitable voltage is applied to the leads, electrons are able to recombine with electron holes within the device, releasing energy in the form of photons. This effect is called electroluminescence, and the color of the light (corresponding to the energy of the photon) is determined by the energy band gap of the semiconductor.

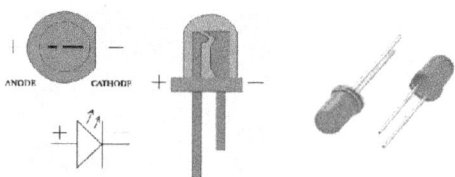

Testing an LED

Never connect an LED directly to a battery or power supply! It will be destroyed almost instantly because too much current will pass through and burn it out.

LEDs must have a resistor in series to limit the current to a safe value, for quick testing purposes a 220 Ω up to 1kΩ resistor are suitable for most LEDs if your supply voltage is 12V or less.

Remember to connect the LED the correct way round!

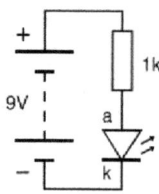

Transistor

A **transistor** is a semiconductor device used to amplify and switch electronic signals and electrical power. It is composed of semiconductor material with at least three terminals for connection to an external circuit. The transistor is called a BJT "Bipolar Junction Transistor"

There are two types of standard transistors, **NPN** and **PNP**, with different circuit symbols. The letters refer to the layers of semiconductor material used to make the transistor. Most transistors used today are NPN because this is the easiest type to make from silicon. This page is mostly about NPN transistors and if you are new to electronics it is best to start by learning how to use these first. The leads are labelled **base** (B), **collector** (C) and **emitter** (E). The arrow inside the transistor symbol indicates the N region.

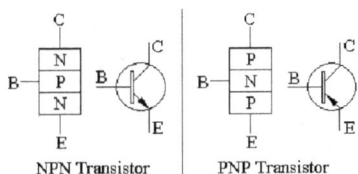

NPN Transistor   PNP Transistor

Using a transistor in the 'switching' configuration is indispensable in many situation, where it is required to drive a relatively important load, providing important currents, that most controllers are not able to provide

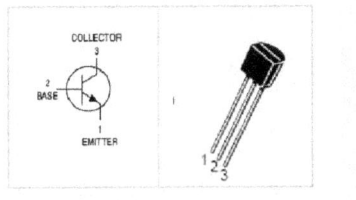 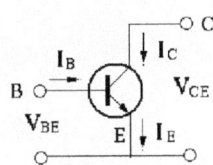

Transistor operation

Transistor OFF: $V_{BE} < 0.6$ V, $I_B \approx 0$

Transistor ON: $V_{BE} >= 0.6$V, $I_B > 0$ $I_C = \beta \, I_B$

Example: $\beta = 200$, $I_B = 0.5$ mA $I_C = 200 \times 0.5 = 100$ mA

BJT Transistor as a Switch

Operating Regions

The areas of operation for a transistor switch are known as the **Saturation Region** and the **Cut-off Region**. This means then that we can ignore the operating Q-point biasing and voltage divider circuitry required for amplification, and use the transistor as a switch by driving it back and forth between its "fully-OFF" (cut-off) and "fully-ON" (saturation) regions as shown below.

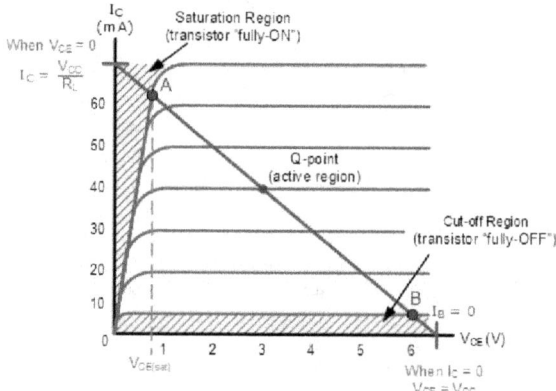

The pink shaded area at the bottom of the curves represents the "Cut-off" region while the blue area to the left represents the "Saturation" region of the transistor. Both these transistor regions are defined as:

1. Cut-off Region

Here the operating conditions of the transistor are zero input base current (I_B), zero output collector current (I_C) and maximum collector voltage (V_{CE}) which results in a large depletion layer and no current flowing through the device. Therefore the transistor is switched "Fully-OFF".

Cut-off Characteristics

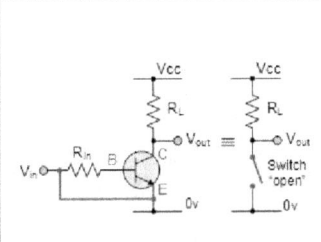

- The input and Base are grounded (0v)
- Base-Emitter voltage $V_{BE} < 0.7v$
- Base-Emitter junction is reverse biased
- Base-Collector junction is reverse biased
- Transistor is "fully-OFF" (Cut-off region)
- No Collector current flows ($I_C = 0$)
- $V_{OUT} = V_{CE} = V_{CC} - I_C R_L = V_{CC}$
- Transistor operates as an "open switch"

2. Saturation Region

Here the transistor will be biased so that the maximum amount of base current is applied, resulting in maximum collector current resulting in the minimum collector emitter voltage drop which results in the depletion layer being as small as possible and maximum current flowing through the transistor. Therefore the transistor is switched "Fully-ON".

Saturation Characteristics

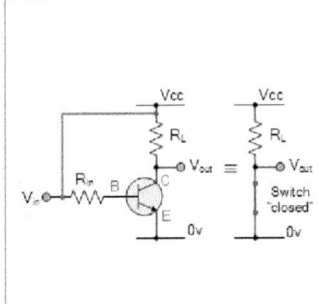

- The input and Base are connected to V_{CC}
- Base-Emitter voltage $V_{BE} > 0.7v$
- Base-Emitter junction is forward biased
- Base-Collector junction is forward biased
- Transistor is "fully-ON" (saturation region)
- Max Collector current flows ($I_C = V_{CC}/R_L$)
- $V_{CE} = 0$ (ideal saturation)
- $V_{OUT} = V_{CE} = "0"$
- Transistor operates as a "closed switch"

Basic NPN Transistor Switching Circuit

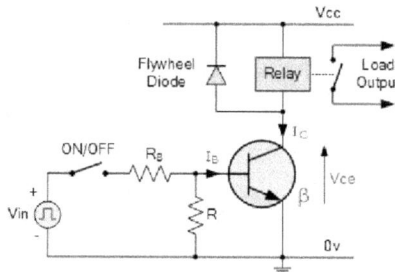

Switch OFF:
....
....
....
....
....

Switch ON
....
....
....
....
....

EEL 2003   Electrical / Electronics Department   SO1-1

Power Electronics

1- Definition of Power Electronics

Power electronics refers to control and conversion of electrical power by power semiconductor devices wherein these devices operate as switches. Advent of silicon-controlled rectifiers, abbreviated as SCRs, led to the development of a new area of application called the power electronics. Once the SCRs were available, the application area spread to many fields such as drives, power supplies, aviation electronics, high frequency inverters and power electronics originated.

Power electronics has applications that span the whole field of electrical power systems, with the power range of these applications extending from a few VA/Watts to several MVA / MW.

In the broadest sense, the task of power electronics is to process and control the flow of electric energy by supplying voltages and currents in a form that is optimally suited for user loads. Fig1.1

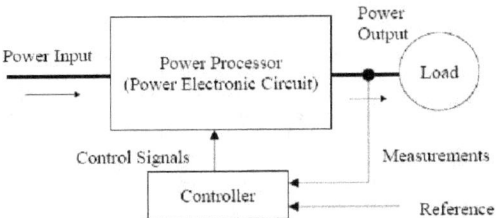

Fig.1.1 Power electronics process-control

2- Converter Classification

The objective of a power electronics circuit is to match the voltage and current requirements of the load to those of the source. Power electronics circuits convert one type or level of a voltage or current waveform to another and are hence called *converters*. Converters serve as an interface between the source and load (Fig. 1-1).

| EEL 2003 | Electrical / Electronics Department | SO1-1 |

"Electronic power converter" is the term that is used to refer to a power electronic circuit that converts voltage and current from one form to another. These converters can be classified as:

- Rectifier converting an AC voltage to a DC voltage,
- Inverter converting a DC voltage to an AC voltage,
- Chopper or a switch-mode power supply that converts a DC voltage to another DC.
- Cycloconverter and cycloinverter converting an AC voltage to another AC voltage.

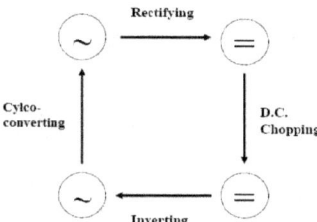

- **ac input/dc output**

The ac-dc converter produces a dc output from an ac input. Average power is transferred from an ac source to a dc load. The ac-dc converter is specifically classified as a *rectifier*. For example, an ac-dc converter enables integrated circuits to operate from a 60-Hz ac line voltage by converting the ac signal to a dc signal of the appropriate voltage.

- **dc input/ac output**

The dc-ac converter is specifically classified as an *inverter*. In the inverter, average power flows from the dc side to the ac side. Examples of inverter applications include producing a 120-V rms 60-Hz voltage from a 12-V battery and interfacing an alternative energy source such as an array of solar cells to an electric utility.

- **dc input/dc output**

The dc-dc converter is useful when a load requires a specified (often regulated) dc voltage or current but the source is at a different or unregulated dc value. For example, 5 V may be obtained from a 12-V source via a dc-dc converter.

- **ac input/ac output**

The ac-ac converter may be used to change the level and/or frequency of an ac signal. Examples include a common light-dimmer circuit and speed control of an induction motor.

3- Semiconductors Switch Types

At this point it is beneficial to review the current state of semiconductor devices used for high power applications. This is required because the operation of many power electronic circuits is intimately tied to the behavior of various devices.

3-1 Power Diodes

The diodes ON and OFF states are controlled by power circuits on the basis of their recovery time ratings. The power diodes are classified into three types+

- General purpose power diodes or conventional diode
- Fast-recovery diodes
- Shottky power diodes

a/ General purpose power diodes

The name itself reveals that these types of power diodes are used for general purposes such as battery charging, UPS and electric traction systems. These power diodes have a high reverse recovery time of about 25us. The availability range of these diodes is current rating of 1A to 1000A and voltage rating of 50V to 5kV.

b/ Fast Recovery Power Diodes

The fast recovery power diodes came into existence for their use in the high frequency switching circuits systems. These are having recovery time of 5ms to 50ms. These are used more than the other diodes in power electronic types. But there are some difficulties in the manufacturing process. The design for these diodes has a voltage rating below 400V.

c/ Schottky Power Diodes

These types of power diodes are usually used in low voltage high frequency applications such as high frequency instrumentation. These diodes have a very fast recovery time compared to the general purpose and fast recovery diodes. The current ratings are in the range of 1A to 300A and reverse voltage rating of about 100V. From the viewpoint of design, it is somewhat different from other diodes. It uses a metal like golden silver platinum on one side of junction and doped silicon (Si) on the other side of junction

Schottky diode schematic symbol

https://www.youtube.com/watch?v=bXEyCf1P0UU

3.2 / Connecting Diodes in Series

Another application for the diode is to create a regulated voltage supply. Diodes are connected together in series to provide a constant DC voltage across the diode combination. The output voltage across the diodes remains constant in spite of changes in the load current drawn from the series combination or changes in the DC power supply voltage that feeds them. Consider the circuit below.

As the forward voltage drop across a silicon diode is almost constant at about 0.7v, while the current through it varies by relatively large amounts, a forward-biased diode can make a simple voltage regulating circuit. The individual voltage drops across each diode are subtracted from the supply voltage to leave a certain voltage potential across the load resistor, and in our simple example above this is given as **10v − (3 x 0.7v) = 7.9v.**

By adding more diodes in series a greater voltage reduction will occur. Also series connected diodes can be placed in parallel with the load resistor to act as a voltage regulating circuit. Here the voltage applied to the load resistor will be **3 x 0.7v = 2.1v.** We can of course produce the same constant voltage source using a single **Zener Diode**. Resistor, R_D is used to prevent excessive current flowing through the diodes if the load is removed.

3.3 / Connecting Diodes in Parallel

In high power applications, diodes are connected in parallel to increase the current carrying capability in order to meet circuit requirements. In parallel operation of diodes, current sharing depends on the magnitude of their forward voltage drops.

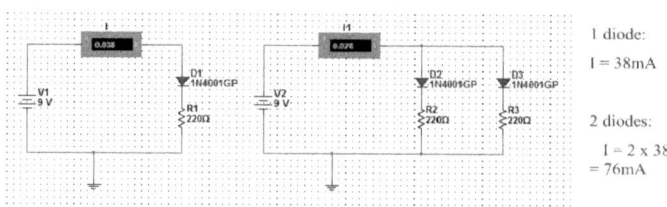

1 diode:
I = 38mA

2 diodes:
I = 2 x 38
= 76mA

3.4 / Freewheel Diodes

Diodes can also be used in a variety of clamping, protection and wave shaping circuits with the most common form of clamping diode circuit being one which uses a diode connected in parallel with a coil or inductive load to prevent damage to the delicate switching circuit by suppressing the voltage spikes and/or transients that are generated when the load is suddenly turned "OFF". This type of diode is generally known as a "Free-wheeling Diode", "Flywheel Diode" or **Freewheel diode** as it is more commonly called.

The **Freewheel diode** is used to protect solid state switches such as power transistors and MOSFET's from damage by reverse battery protection as well as protection from highly inductive loads such as relay coils or motors, and an example of its connection is shown below.

Use of the Freewheel Diode

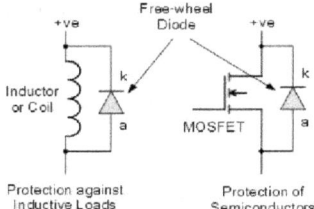

Modern fast switching, power semiconductor devices require fast switching diodes such as free wheeling diodes to protect them form inductive loads such as motor coils or relay windings. Every time the switching device above is turned "ON", the freewheel diode changes from a conducting state to a blocking state as it becomes reversed biased.

However, when the device rapidly turns "OFF", the diode becomes forward biased and the collapse of the energy stored in the coil causes a current to flow through the freewheel diode. Without the protection of the freewheel diode high di/dt currents would occur causing a high voltage spike or transient to flow around the circuit possibly damaging the switching device.

Previously, the operating speed of the semiconductor switching device, either transistor, MOSFET, IGBT or digital has been impaired by the addition of a freewheel diode across the inductive load with Schottky and Zener diodes being used instead in some applications. But during the past few years however, freewheel diodes had regained importance due mainly to their improved reverse-recovery characteristics and the use of super fast semiconductor materials capable at operating at high switching frequencies.

https://www.youtube.com/watch?v=LXGtE3X2k7Y

| EEL 2003 | Electrical / Electronics Department | SO1/2 |

Thyristor

In many ways the Silicon Controlled Rectifier, or the Thyristor as it is more commonly known, is similar to the transistor. It is a multi-layer semiconductor device, hence the "silicon" part of its name. It requires a gate signal to turn it "ON", the "controlled" part of the name and once "ON" it behaves like a rectifying diode, the "rectifier" part of the name Figure 1. In fact the circuit symbol for the *thyristor* suggests that this device acts like a controlled rectifying diode.

A **thyristor** is a solid-state semiconductor device with four layers of alternating N and P-type material. It acts exclusively as a bistable switch, conducting when the gate receives a current trigger, and continuing to conduct while the voltage across the device is not reversed (forward-biased).

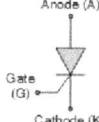

Figure 1: Thyristor Symbol

1- Thyristor representation

However, unlike the diode which is a two layer (P-N) semiconductor device, or the transistor which is a three layer (P-N-P, or N-P-N) device, the **Thyristor** is a four layer (P-N-P-N) semiconductor device that contains three PN junctions in series, and is represented by the symbol as shown in figure 2.

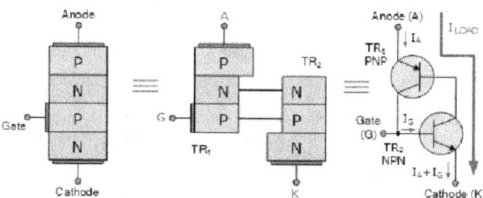

Figure 2: Thyristor representation

Like the diode, the Thyristor is a unidirectional device, that is it will only conduct current in one direction only, but unlike a diode, the thyristor can be made to operate as either an open-circuit switch or as a rectifying diode depending upon how the thyristors gate is triggered. In other words, thyristors can operate only in the switching mode and cannot be used for amplification.

The silicon controlled rectifier **SCR**, is one of several power semiconductor devices along with Triacs (Triode AC's), Diacs (Diode AC's) and UJT's (Unijunction Transistor) that are all capable of acting like very fast solid state AC switches for controlling large AC voltages and currents. So for the Electronics student this makes these very handy solid state devices for controlling AC motors, lamps and for phase control.

The thyristor is a three-terminal device labelled: "Anode", "Cathode" and "Gate" and consisting of three PN junctions which can be switched "ON" and "OFF" at an extremely fast rate, or it can be switched "ON" for variable lengths of time during half cycles to deliver a selected amount of power to a load. The operation of the thyristor can be best explained by assuming it to be made up of two transistors connected back-to-back as a pair of complementary regenerative switches as shown in figure 3.

2- Thyristor I-V Characteristics Curves

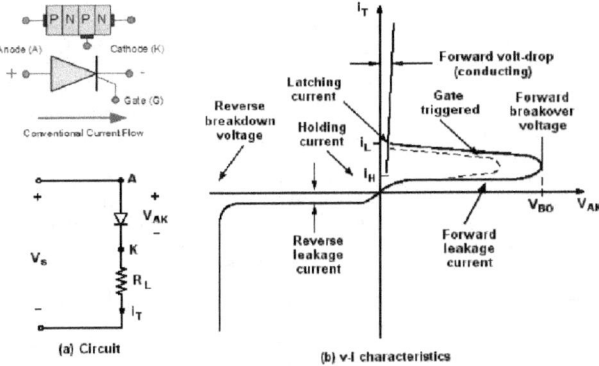

Figure 3: Thyristor Characteristics

The important points on this characteristic are:

Latching Current I_L

This is the minimum anode current required to maintain the thyristor in the on-state immediately after a thyristor has been turned on and the gate signal has been removed. If a gate current greater than the threshold gate current is applied until the anode current is greater than the latching current I_L then the thyristor will be turned on or triggered.

Holding Current I_H

This is the minimum anode current required to maintain the thyristor in the on-state. To turn off a thyristor, the forward anode current must be reduced below its holding current for a sufficient time for mobile charge carriers to vacate the junction. If the anode current is not maintained below I_H for long enough, the thyristor will not have returned to the fully blocking state by the time the anode-to-cathode voltage rises again. It might then return to the conducting state without an externally-applied gate current.

Reverse Current I_R

When the cathode voltage is positive with respect to the anode, the junction J_2 is forward biased but junctions J_1 and J_3 are reverse biased. The thyristor is said to be in the *reverse blocking state* and a reverse leakage current known as reverse current I_R will flow through the device.

Forward Breakover Voltage V_{BO}

If the forward voltage V_{AK} is increased beyond V_{BO}, the thyristor can be turned on. But such a turn-on could be destructive. In practice the forward voltage is maintained below V_{BO} and the thyristor is turned on by applying a positive gate signal between gate and cathode.

Once the thyristor is turned on by a gate signal and its anode current is greater than the holding current, the device continues to conduct due to positive feedback even if the gate signal is removed. This is because the thyristor is a latching device and it has been latched to the on-state.

So how do we turn "OFF" the thyristor?. Once the thyristor has self-latched into its "ON" state and passing a current, it can only be turned "OFF" again by either removing the supply voltage and therefore the Anode (I_A) current completely, or by reducing its Anode to Cathode current by some external means (the opening of a switch for example) to below a value commonly called the "minimum holding current", I_H.

3- Turn ON and OFF methods of SCR

3-1 Turn on methods of an SCR

the SCR can be switched on either by increasing the forward voltage beyond forward break over voltage VFB0 or by applying a positive gate signal when the device is forward biased. Of these two methods, the latter, called the gate-control method, is used as it is more efficient and easy to implement for power control.

The following points have to be noted when designing the gate-control circuit

1. Appropriate gate-to-cathode voltage must be applied for turn-on when the device is forward biase
2. The gate signal must be removed after the device is turned-on
3. No gate signal should be applied when the device is reverse-biased

The most used methods for triggering an SCR or thyristor are mentioned below:

- Voltage Triggering
- Gate triggering

a. Voltage Triggering

The method of triggering in which the triggering of SCR or thyristor is caused by the applied voltage across anode and cathode terminals, is known as Voltage triggering.

When a thyristor is in forward biased and applied voltage across anode and cathode is increased, then the depletion layer of reverse biased junction decreased. At breakdown voltage, the depletion layer is totally destroyed and as a result the thyristor triggers and comes to ON state and start conducting heavily due to increase in number of charge carriers.

b. Gate triggering

The triggering method in which triggering of SCR or thyristor is caused by applying a signal between gate and cathode, is called Gate triggering.
Gate triggering is mostly used method for triggering an SCR or thyristor. This method is used in almost all industries and laboratories to trigger thyristor. In this method a positive signal is applied in between gate and cathode terminal.

By using this method we can trigger the device much before its breakdown voltage. Hence, we can also control the firing angle (α) and the conduction angle ($\beta=180-\alpha$) of SCR or thyristor.

Example of gate triggering

At the start of each positive half-cycle the SCR is "OFF". On the application of the gate pulse triggers the SCR into conduction and remains fully latched "ON" for the duration of the positive cycle. If the thyristor is triggered at the beginning of the half-cycle ($\theta = 0°$), the load (a lamp) will be "ON" for the full positive cycle of the AC waveform (half-wave rectified AC).

As the application of the gate trigger pulse increases along the half cycle ($\theta = 0°$ to $90°$), the lamp is illuminated for less time and the average voltage delivered to the lamp will also be proportionally less reducing its brightness.

Then we can use a silicon controlled rectifier as an AC light dimmer as well as in a variety of other AC power applications such as: AC motor-speed control, temperature control systems and power regulator circuits, etc.

Thus far we have seen that a thyristor is essentially a half-wave device that conducts in only the positive half of the cycle when the Anode is positive and blocks current flow like a diode when the Anode is negative, irrespective of the Gate signal.

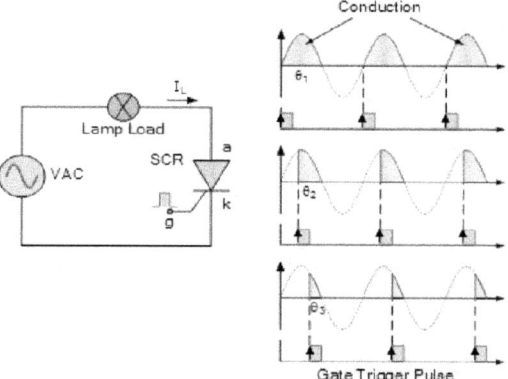

Figure 4: Triggering the thyristor in AC

By varying the time of the triggering pulse, the current flowing through the load (bulb) will be varied from zero to its maximum value.

3-2 Methods of Turning Off the thyristor

It is recommended to know about SCR-Basics, Structure, Characteristics before proceeding further, short duration pulse is applied to the gate (typically 4V, 100μs).

Once the thyristor is turned-on, the gate loses control and the thyristor will only turn off when the load current falls virtually to zero, or the thyristor is reverse biased.

The thyristor will turn off naturally with a.c. supplies as the voltage reverses (which is called as Natural Commutation), but no such reversal occurs with d.c. supplies and it is necessary to force a voltage reversal if turn-off is to occur. This process is called **Forced Commutation**.

A- Commutation:

The process of turning OFF SCR is defined as "Commutation".

- In all commutation techniques, a reverse voltage is applied across the thyristor during the turn OFF process.
- By turning OFF a thyristor we bring it from forward conducting to the forward blocking mode.

The condition to be satisfied in order to turn Off an SCR are:

a- $I_A < I_H$ (Anode current must be less than holding current)
b- A reverse voltage is applied to SCR for sufficient time enabling it to recover its blocking state.

- There are two methods by which a thyristor can be turned Off.

   a- Natural Commutation
   b- Forced Commutation

B- Natural Commutation:

- In AC circuit, the current always passes through zero for every half cycle as shown in figure 4
- As the current passes through natural zero, a reverse Voltage will simultaneously appear across the device.
- This will turn OFF the device immediately.
- This process is called as natural commutation, since no external circuit is required for this purpose.

EEL 2003   Electrical / Electronics Department   SO1/2

C- Forced Commutation:

- To turn OFF a thyristor, the forward anode current should be brought to zero for sufficient time to allow the removal of charged carriers.
- In case of DC circuits the forward current should be forced to zero by means of some external circuits as shown in figure 5
- This process is called as forced commutation

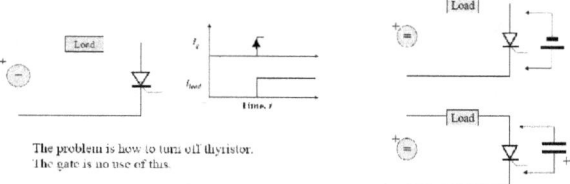

Figure 5: Turning On / Off the thyristor in DC

a- Thyristor Types:

There are many types of thyristors all of them has three terminals but differs only in how they can turn ON and OFF. The most famous types of thyristors are:

1. Silicon controlled rectifier (SCR)
3. Gate-turn-off thyristor (GTO)
4. Bidirectional triode thyristor (TRIAC)
5. Light activated silicon-controlled rectifier (LASCR)

The electric circuit symbols of each type of thyristors are shown in Fig.7

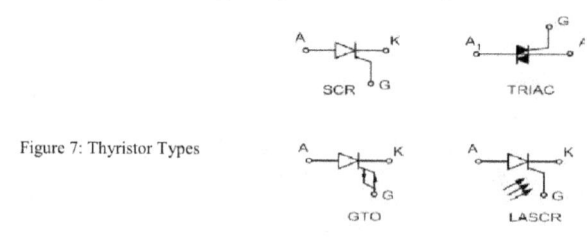

Figure 7: Thyristor Types

- **Gate Turn Off thyristor (GTO).**

A GTO thyristor can be turned on by a single pulse of positive gate current like conventional thyristor, but in addition it can be turned off by a pulse of negative gate current as shown in figure 8. The gate current therefore controls both ON state and OFF state operation of the device. The GTO has many advantages and disadvantages with respect to conventional thyristor here will talk about these advantages and disadvantages.

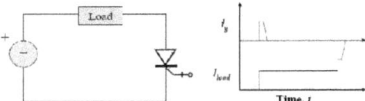

Figure 8: Operation of the GTO

The GTO has the following advantage over thyristor:

1- Elimination of commutating components in forced commutation resulting in reduction in cost, weight and volume,
2- Faster turn OFF permitting high switching frequency,
3- Improved converters efficiency, and,
4- It has more di/dt rating at turn ON.

The thyristor has the following advantage over GTO.

1- ON state voltage drop and associated losses are higher in GTO than thyristor,
2- Triggering gate current required for GTOs is more than those of thyristor,
3- Latching and holding current is more in GTO than those of thyristor,
3- Gate drive circuit loss is more than those of thyristor, and
4- Its reverse voltage block capability is less than its forward blocking capability.

- **Bi-Directional Thyristor (TRIAC).**

TRIAC are used for the control of power in AC circuits. A TRIAC is equivalent of two reverse parallel-connected SCRs with one common gate. Conduction can be achieved in either direction with an appropriate gate current. A TRIAC is thus a bi-directional gate controlled thyristor with three terminals. The terms anode and cathode are not applicable to TRIAC. Fig.9 shows the i-v characteristics of the TRIAC.

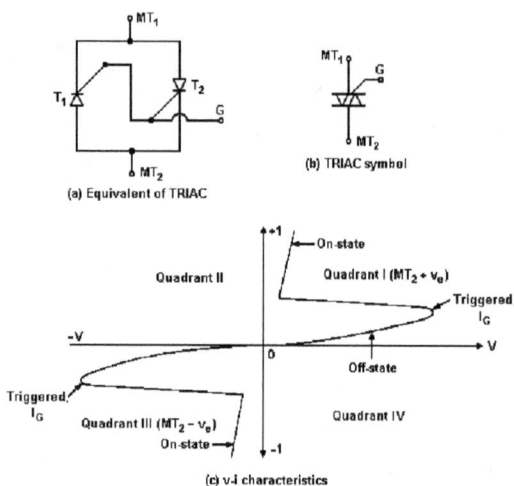

Figure 9: Triac Characteristic

- **The Diac**

This is a bi-directional trigger diode used mainly in firing Triacs and Thyristors in AC control circuits. Its circuit symbol (shown in figure 3a) is similar to that of a Triac, but without the gate terminal, in fact it is a simpler device and consists of a PNP structure (like a transistor without a base) and acts basically as two diodes connected cathode to cathode as shown in figure 3b.

Figure 10. The Diac Circuit symbol and an equivalent diagram using diodes.

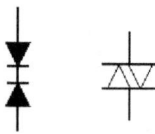

The DIAC is designed to have a particular break over voltage, typically about 30 volts, and when a voltage less than this is applied in either polarity, the device remains in a high resistance state with only a small leakage current flowing.

Once the break over voltage is reached however, in either polarity, the device exhibits a negative resistance as can be seen from the characteristic curve in Figure 4.

- **Typical Diac Characteristics.**

When the voltage across the diac exceeds about 30 volts (a typical break-over voltage) current flows and an increase in current is accompanied by a drop in the voltage across the Diac. Normally, Ohm's law states that an increase in current through a component causes an increase in voltage across that component; however the opposite effect is happening here, therefore the Diac exhibits negative resistance at break-over.

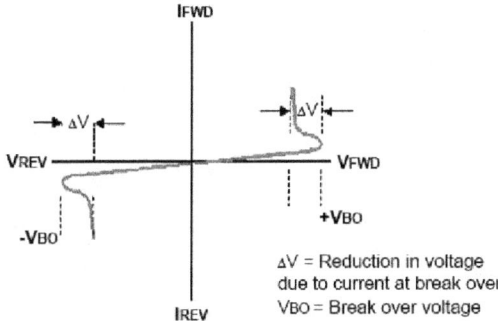

- **Typical Diac and Triac Application**

Triacs are widely used in applications such as lamp dimmers and motor speed controllers and as such the diac is used in conjunction with the triac to provide full-wave control of the AC supply as shown. The diac is commonly used as a triggering device for other semiconductor switching devices, mainly SCR's and triacs.

The figure 10 shows a Quadrac which is a special type of thyristor which combines a "**diac**" and a "triac" in a single package. The **diac** is the triggering device for the triac.

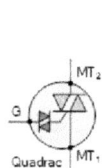
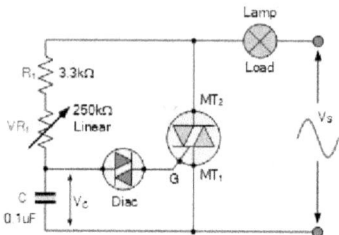

Figure 10: Quadrac   Figure 11: Triac controlled by a Diac

As the AC supply voltage increases at the beginning of the cycle, capacitor, C is charged through the series combination of the fixed resistor, R1 and the potentiometer, VR1 and the voltage across its plates increases. When the charging voltage reaches the breakover voltage of the diac (about 30 V), the diac breaks down and the capacitor discharges through the diac, producing a sudden pulse of current, which fires the triac into conduction. The phase angle at which the triac is triggered can be varied using VR1, which controls the charging rate of the capacitor.

Once the triac has been fired into conduction, it is maintained in its "ON" state by the load current flowing through it, while the voltage across the resistor–capacitor combination is limited by the "ON" voltage of the triac and is maintained until the end of the present half-cycle of the AC supply.

At the end of the half cycle the supply voltage falls to zero, reducing the current through the triac below its holding current, I_H turning it "OFF" and the diac stops conduction. The supply voltage then enters its next half-cycle, the capacitor voltage again begins to rise (this time in the opposite direction) and the cycle of firing the triac repeats over again.

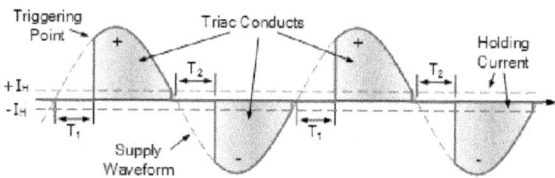

Figure 12: Triac Conduction waveform

4. Thyristor Protection

- For reliable operation of SCR, it should be operated within the specific ratings.
- SCRs are very delicate devices and so they must be protected against abnormal operating conditions. Two of the various protection of SCR are

    a- di/dt Protection
    b- dv/dt Protection

A- High dv/dt Protection

When a thyristor is in forward blocking state then only J_2 junction is reverse biased which acts as a capacitor having constant capacitance value C_j (junction capacitance). As we know that current through capacitor follows the relation

$$i = C\frac{dv}{dt} \Rightarrow i \propto \frac{dv}{dt} (if\ C\ constant)$$

Hence leakage current through the J_2 junction which is nothing but the leakage current through the device will increase with the increase in dv_a/dt i.e. rate of change of applied voltage across the thyristor. This current can turn-on the device even when the gate signal is absent. This is called dv/dt triggering and must be avoided which can be achieved by using Snubber circuit in parallel with the device.

Protective Measure:

Snubber Circuit:

It consists of a capacitor connected in series with a resistor which is applied parallel with the thyristor, when S is closed then voltage V_s is applied across the device as well as C_s suddenly. At first Snubber circuit behaves like a short circuit. Therefore voltage across the device is zero. Gradually voltage across C_s builds up at a slow rate. So dv/dt across the thyristor will stay in allowable range.

Before turning on of thyristor C_s is fully charged and after turning on of thyristor it discharges through the SCR. This discharging current can be limited with the help of a resistance (R_s) connected in series with the capacitor (C_s) to keep the value of current and rate of change of current in a safe limit.

EEL 2003　　　　Electrical / Electronics Department　　　　SO1/2

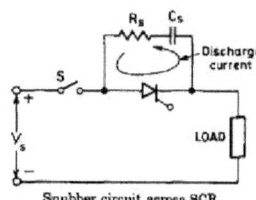

Snubber circuit across SCR.

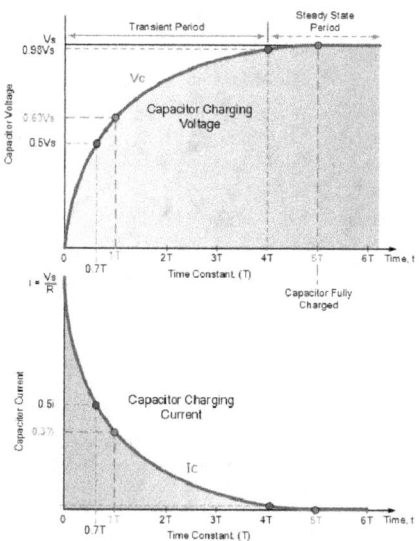

B- High di/dt Protection

When a thyristor is turned on by **gate pulse** then charge carriers spread through its junction rapidly. But if rate of rise of anode current, i.e. di/dt is greater than the spreading of charge carriers then localized heat generation will take place which is known as local hot spots. This may damage the thyristor.

Protective Measure:

To avoid local hot spots we use an inductor in series with the device as it prevents high rate of change of current through it.
Local spot heating is avoided by ensuring that the conduction spreads to the whole area very rapidly. (OR) The di/dt value must be maintained below a threshold (limiting) value.
This is done by means of connecting an inductor in series with the thyristor.
The inductance L opposes the high di/dt variations.
When the current variation is high, the inductor smooths it and protects the SCR from damage. (Though di/dt variation is high, the inductor 'L' smooths it because it takes some time to charge).
$L \geq [V_s / (di/dt)]$.

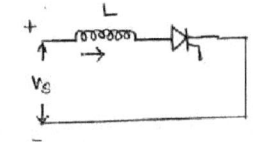

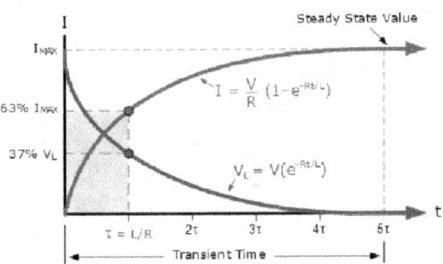

Thyristor Switch Applications

DC Thyristor Circuit

When connected to a direct current DC supply, the thyristor can be used as a DC switch to control larger DC currents and loads. When using the Thyristor as a switch it behaves like an electronic latch because once activated it remains in the "ON" state until manually reset.

Consider the DC thyristor circuit in figure below

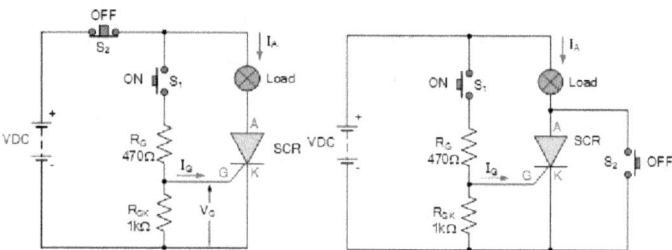

This simple "on-off" thyristor firing circuit uses the thyristor as a switch to control a lamp, but it could also be used as an on-off control circuit for a motor, heater or some other such DC load.

The thyristor is forward biased and is triggered into conduction by briefly closing the normally-open "ON" push button, S_1 which connects the Gate terminal to the DC supply via the Gate resistor, R_G thus allowing current to flow into the Gate. If the value of R_G is set too high with respect to the supply voltage, the thyristor may not trigger.

Once the circuit has been turned-"ON", it self latches and stays "ON" even when the push button is released providing the load current is more than the thyristors latching current. Additional operations of push button, S_1 will have no effect on the circuits state as once "latched" the Gate looses all control. The thyristor is now turned fully "ON" (conducting) allowing full load circuit current to flow through the device in the forward direction and back to the battery supply.

One of the main advantages of using a thyristor as a switch in a DC circuit is that it has a very high current gain. The thyristor is a *current operated device* because a small Gate current can control a much larger Anode current.

The Gate-cathode resistor R_{GK} is generally included to reduce the Gate's sensitivity and increase its dv/dt capability thus preventing false triggering of the device.

As the thyristor has self latched into the "ON" state, the circuit can only be reset by interrupting the power supply and reducing the Anode current to below the thyristors minimum holding current (I_H) value.

Opening the normally-closed "OFF" push button, S_2 breaks the circuit, reducing the circuit current flowing through the **Thyristor** to zero, thus forcing it to turn "OFF" until the application again of another Gate signal.

However, one of the disadvantages of this DC thyristor circuit design is that the mechanical normally-closed "OFF" switch S_2 needs to be big enough to handle the circuit power flowing through both the thyristor and the lamp when the contacts are opened. If this is the case we could just replace the thyristor with a large mechanical switch. One way to overcome this problem and reduce the need for a larger more robust "OFF" switch is to connect the switch in parallel with the thyristor as shown.

Here the thyristor switch receives the required terminal voltage and Gate pulse signal as before but the larger normally-closed switch of the previous circuit has be replaced by a smaller normally-open switch in parallel with the thyristor. Activation of switch S_2 momentarily applies a short circuit between the thyristors Anode and Cathode stopping the device from conducting by reducing the holding current to below its minimum value.

AC Thyristor Circuit

When connected to an alternating current AC supply, the thyristor behaves differently from the previous DC connected circuit. This is because AC power reverses polarity periodically and therefore any thyristor used in an AC circuit will automatically be reverse-biased causing it to turn-"OFF" during one-half of each cycle.

Consider the AC thyristor circuit in figure 4 below.

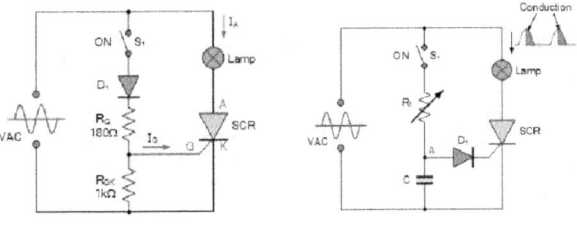

The above thyristor firing circuit is similar in design to the DC SCR circuit except for the omission of an additional "OFF" switch and the inclusion of diode D_1 which prevents reverse bias being applied to the Gate. During the positive half-cycle of the sinusoidal waveform, the device is forward biased but with switch S_1 open, zero gate current is applied to the thyristor and it remains "OFF". On the negative half-cycle, the device is reverse biased and will remain "OFF" regardless of the condition of switch S_1.

If switch S_1 is closed, at the beginning of each positive half-cycle the thyristor is fully "OFF" but shortly after there will be sufficient positive trigger voltage and therefore current present at the Gate to turn the thyristor and the lamp "ON".

The thyristor is now latched-"ON" for the duration of the positive half-cycle and will automatically turn "OFF" again when the positive half-cycle ends and the Anode current falls below the holding current value.

During the next negative half-cycle the device is fully "OFF" anyway until the following positive half-cycle when the process repeats itself and the thyristor conducts again as long as the switch is closed.

Then in this condition the lamp will receive only half of the available power from the AC source as the thyristor acts like a rectifying diode, and conducts current only during the positive half-cycles when it is forward biased. The thyristor continues to supply half power to the lamp until the switch is opened.

If it were possible to rapidly turn switch S_1 ON and OFF, so that the thyristor received its Gate signal at the "peak" (90°) point of each positive half-cycle, the device would only conduct for one half of the positive half-cycle. In other words, conduction would only take place during one-half of one-half of a sine wave and this condition would cause the lamp to receive "one-fourth" or a quarter of the total power available from the AC source.

By accurately varying the timing relationship between the Gate pulse and the positive half-cycle, the **Thyristor** could be made to supply any percentage of power desired to the load, between 0% and 50%. Obviously, using this circuit configuration it cannot supply more than 50% power to the lamp, because it cannot conduct during the negative half-cycles when it is reverse biased.

Consider the circuit below.

Phase control is the most common form of thyristor AC power control and a basic AC phase-control circuit can be constructed as shown above. Here the thyristors Gate voltage is derived from the RC charging circuit via the trigger diode, D_1.

During the positive half-cycle when the thyristor is forward biased, capacitor, C charges up via resistor R_1 following the AC supply voltage. The Gate is activated only when the voltage at point A has risen enough to cause the trigger diode D_1, to conduct and the capacitor discharges into the Gate of the thyristor turning it "ON". The time duration in the positive half of the cycle at which conduction starts is controlled by RC time constant set by the variable resistor, R_1.

B. Part 2: Describe firing and driving circuits for power electronic converters.

1. Describe ideal and non-ideal properties of operational amplifiers. Determine the operation of various related circuits (inverting and non-inverting amplifiers, buffer amplifier, summing amplifier)
2. Describe the use of an operational amplifier for PWM generation, for triangular and sine wave generation, as a comparator, and its integration into a 555 timer.
3. Explore other basic firing and driving circuits by focusing on requirements and control features such as based on specific power devices and operational amplifier.

EEL 2003   Electrical / Electronics Department   LO # 2

Operational Amplifiers

1- Introduction: Amplifier Circuit

A part from their most common use as amplifiers (both inverting and non-inverting), they also find applications as buffers (load isolators), adders, subtractors, integrators, logarithmic amplifiers, impedance converters, filters (low-pass, high-pass, band-pass, band-reject or notch), and differential amplifiers.

Before jumping into op-amps, let's first go over some amplifier fundamentals. An amplifier has an *input port* and an *output port*. (A port consists of two terminals, one of which is usually connected to the ground node.)

In a linear amplifier, the **Output signal = A × Input signal,**

where A is the amplification factor or "gain."

Depending on the nature of the input and output signals, we can have four types of amplifier gain:
- Voltage gain (voltage out / voltage in),
- Current gain (current out / current in),
- Transresistance (voltage out / current in)
- Transconductance (current out / voltage in).

Since most op-amps are voltage/voltage amplifiers, we will limit the discussion here to this type of amplifier. The circuit model of an amplifier is shown in Figure 1 (center dashed box, with an input port and an output port).

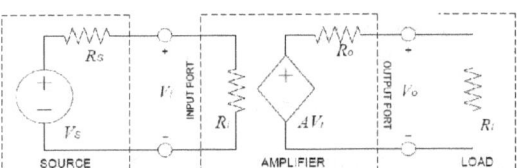

Figure 1: Circuit model of an amplifier circuit.

The input port plays a passive role, producing no voltage of its own, and is modelled by a resistive element R_i called the *input resistance*. The output port is modeled by a dependent voltage source AV_i in series with the *output resistance* R_o, where V_i is the potential difference between the input port terminals.

Figure 1 shows a complete amplifier circuit, which consists of an input voltage source V_s in series with the source resistance R_s, and an output "load" resistance R_L. From this figure, it can be seen that we have voltage-divider circuits at both the input port and the output port of the amplifier. This requires us to re-calculate V_i and V_o whenever a different source and/or load is used:

$$V_i = V_S (R_i / R_S + R_i) \qquad (1)$$

$$V_o = AV_i (R_L / R_o + R_L) \qquad (2)$$

2- The Operational Amplifier: Ideal Op-Amp Model

The amplifier model shown in Figure 1 is redrawn in Figure 2 showing the standard op-amp notation. An op-amp is a "differential to single-ended" amplifier, i.e. it amplifies the voltage difference $V_p - V_n = V_i$ at the input port and produces a voltage V_o at the output port that is referenced to the ground node of the circuit in which the op-amp is used.

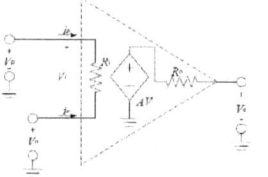
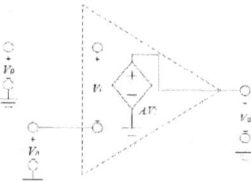

Figure 2: Standard op-amp   Figure 3: Ideal op-amp

The ideal op-amp model was derived to simplify circuit analysis and is commonly used by engineers for first-order approximation calculations. The ideal model makes three simplifying assumptions:

$$\text{Gain is infinite: } A = \infty \qquad (3)$$

$$\text{Input resistance is infinite: } R_i = \infty \qquad (4)$$

$$\text{Output resistance is zero: } R_o = 0 \qquad (5)$$

Applying these assumptions to the standard op-amp model results in the ideal op-amp model shown in Figure 3. Because $R_i = \infty$ and the voltage difference $V_p - V_n = V_i$ at the input port is finite, the input currents are zero for an ideal op-amp:

$$i_n = i_p = 0 \qquad (6)$$

Hence there is no loading effect at the input port of an ideal op-amp. Equation (1) becomes:

$$V_i = V_s \qquad (7)$$

EEL 2003   Electrical / Electronics Department   LO # 2

In addition, because $R_0 = 0$, there is no loading effect at the output port of an ideal op-amp. Equation (2) becomes:

$$V_0 = A \times V_i \qquad (8)$$

Finally, with $A = \infty$, if $V_0 = 0$ then $V_i = V_p - V_n = 0$, or

$$V_p = V_n \qquad (9)$$

Note: Although Equations 3-5 constitute the ideal op-amp assumptions, Equations 6 and 9 are used most often in solving op-amp circuits.

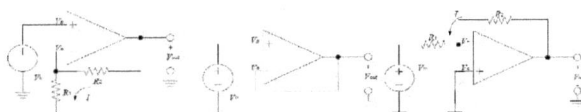

Figure 4a: Non-inverting amplifier   Figure 5a: Voltage follower   Figure 6a: Inverting amplifier

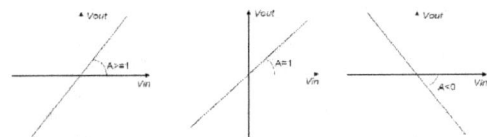

Figure 4b: Voltage transfer curve of non-inverting amplifier   Figure 5b: Voltage transfer curve of voltage follower   Figure 6b: Voltage transfer curve of inverting amplifier

The supply voltage $+V$ and $-V$ limit the output voltage as shown below

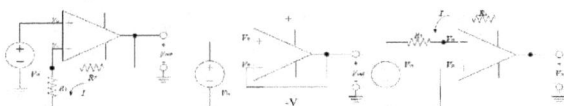

Figure 4a: Non-inverting amplifier   Figure 5a: Voltage follower   Figure 6a: Inverting amplifier

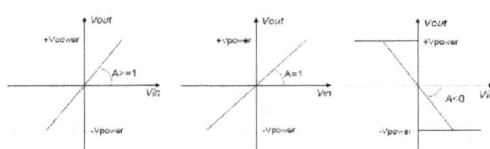

Figure 4c: Realistic transfer curve of non-inverting amplifier   Figure 5c: Realistic transfer curve of voltage follower   Figure 6c: Realistic transfer curve of inverting amplifier

3- The Inverting Operational Amplifier

We saw in the last tutorial that the Open Loop Gain, (Avo) of an operational amplifier can be very high, as much as 1,000,000 or more. However, this very high gain is of no real use to us as it makes the amplifier both unstable and hard to control as the smallest of input signals, just a few micro-volts, (μV) would be enough to cause the output voltage to saturate and swing towards one or the other of the voltage supply rails losing complete control of the output.

As the open loop DC gain of an **Operational Amplifiers** is extremely high we can therefore afford to lose some of this high gain by connecting a suitable resistor across the amplifier from the output terminal back to the inverting input terminal to both reduce and control the overall gain of the amplifier. This then produces and effect known commonly as **Negative Feedback**, and thus produces a very stable Operational Amplifier based system.

Negative Feedback is the process of "feeding back" a fraction of the output signal back to the input, but to make the feedback negative, we must feed it back to the negative or "inverting input" terminal of the op-amp using an external **Feedback Resistor** called Rf. This feedback connection between the output and the inverting input terminal forces the differential input voltage towards zero.

This effect produces a closed loop circuit to the amplifier resulting in the gain of the amplifier now being called its **Closed-loop Gain**. Then a closed-loop inverting amplifier uses negative feedback to accurately control the overall gain of the amplifier, but at a cost in the reduction of the amplifiers gain.

This negative feedback results in the inverting input terminal having a different signal on it than the actual input voltage as it will be the sum of the input voltage plus the negative feedback voltage giving it the label or term of a *Summing Point*. We must therefore separate the real input signal from the inverting input by using an **Input Resistor**, Rin.

As we are not using the positive non-inverting input this is connected to a common ground or zero voltage terminal as shown below, but the effect of this closed loop feedback circuit results in the voltage potential at the inverting input being equal to that at the non-inverting input producing a *Virtual Earth* summing point because it will be at the same potential as the grounded reference input. In other words, the op-amp becomes a "differential amplifier".

Inverting Operational Amplifier Configuration

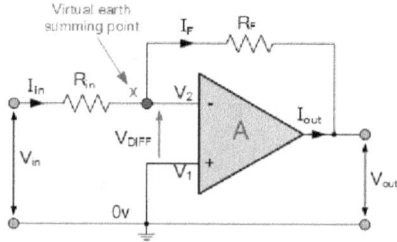

EEL 2003   Electrical / Electronics Department   LO # 2

In this **Inverting Amplifier** circuit the operational amplifier is connected with feedback to produce a closed loop operation. When dealing with operational amplifiers there are two very important rules to remember about inverting amplifiers, these are: **"No current flows into the input terminal"** and that **"V1 always equals V2"**. However, in real world op-amp circuits both of these rules are slightly broken.

This is because the junction of the input and feedback signal (X) is at the same potential as the positive (+) input which is at zero volts or ground then, the junction is a **"Virtual Earth"**. Because of this virtual earth node the input resistance of the amplifier is equal to the value of the input resistor, Rin and the closed loop gain of the inverting amplifier can be set by the ratio of the two external resistors.

We said above that there are two very important rules to remember about **Inverting Amplifiers** or any operational amplifier for that matter and these are.

- **1. No Current Flows into the Input Terminals**
- **2. The Differential Input Voltage is Zero as V1 = V2 = 0** (Virtual Earth)

Then by using these two rules we can derive the equation for calculating the closed-loop gain of an inverting amplifier, using first principles.

Current (i) flows through the resistor network as shown.

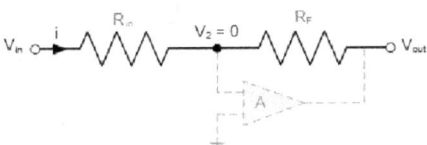

Gain (AV) = V_{OUT} / V_{IN} = **- R_F / R_{IN}**

Demonstration:

V1 = 0V (Grounded), then V2 = V1 = 0V

$V_{IN} = R_{IN} \times I_{IN}$ and $V_{OUT} = - R_F \times I_F$

$I_{IN} = I_F + I_n$ ($I_n = 0$), thus $I_{IN} = I_F$

$V_{OUT} / V_{IN} = (- R_F \times I_F) / (R_{IN} \times I_{IN}) = - R_F / R_{IN}$

Then, the **Closed-Loop Voltage Gain** of an Inverting Amplifier is given as.

$$\text{Gain}(Av) = \frac{V_{out}}{V_{in}} = -\frac{R_f}{R_{in}}$$

and this can be transposed to give Vout as:

$$V_{out} = -\frac{Rf}{Rin} \times Vin$$

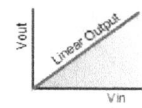

Linear Output

The sign minus (-) indicates that the voltage output in being inverted because the V_{in} is connected to minus terminal of the opamp via the resistance R_{in}.

Inverting Op-amp Example No1

Find the closed loop gain of the following inverting amplifier circuit.

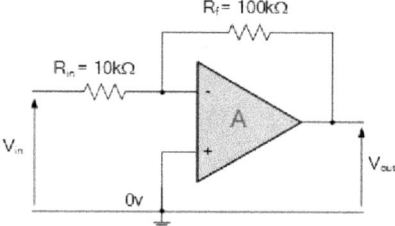

Using the previously found formula for the gain of the circuit

$$\text{Gain}(Av) = \frac{V_{out}}{V_{in}} = -\frac{R_f}{R_{in}}$$

We can now substitute the values of the resistors in the circuit as follows,

Rin = 10kΩ and Rf = 100kΩ.

and the gain of the circuit is calculated as   **-Rƒ/Rin = 100k/10k = -10.**

therefore, the closed loop gain of the inverting amplifier circuit above is given **-10**

Inverting Op-amp Example No2

The gain of the original circuit is to be increased to **40**, find the new values of the resistors required.

Assume that the input resistor is to remain at the same value of 10KΩ, then by re-arranging the closed loop voltage gain formula we can find the new value required for the feedback resistor Rf.

$$\text{Gain} = Rf/Rin$$

therefore, Rf = Gain x Rin

$$Rf = 40 \times 10{,}000$$

$$Rf = 400{,}000 \text{ or } 400\text{K}\Omega$$

The new values of resistors required for the circuit to have a gain of **40** would be,

$$Rin = 10\text{K}\Omega \text{ and } Rf = 400\text{K}\Omega.$$

The formula could also be rearranged to give a new value of Rin, keeping the same value of Rf.

One final point to note about the **Inverting Amplifier** configuration for an operational amplifier, if the two resistors are of equal value, Rin = Rf then the gain of the amplifier will be -1 producing a complementary form of the input voltage at its output as Vout = -Vin. This type of inverting amplifier configuration is generally called a **Unity Gain Inverter** of simply an *Inverting Buffer*.

4- The Non-inverting Operational Amplifier

The second basic configuration of an operational amplifier circuit is that of a Non-inverting Operational Amplifier. In this configuration, the input voltage signal, (Vin) is applied directly to the non-inverting (+) input terminal which means that the output gain of the amplifier becomes "Positive" in value in contrast to the "Inverting Amplifier" circuit we saw in the last tutorial whose output gain is negative in value. The result of this is that the output signal is "in-phase" with the input signal.

Feedback control of the **Non-inverting Operational Amplifier**

Loading product data is achieved by applying a small part of the output voltage signal back to the inverting (−) input terminal via a Rf − R2 voltage divider network, again producing negative feedback. This closed-loop configuration produces a non-inverting amplifier circuit with very good stability, a very high input impedance, Rin approaching infinity, as no current flows into the positive input terminal, (ideal conditions) and a low output impedance, Rout as shown below.

EEL 2003   Electrical / Electronics Department   LO # 2

Non-inverting Operational Amplifier Configuration

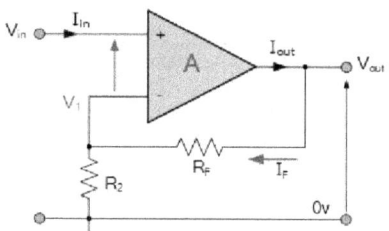

In the previous **Inverting Amplifier** tutorial, we said that for an ideal op-amp **"No current flows into the input terminal"** of the amplifier and that **"V1 always equals V2"**. This was because the junction of the input and feedback signal (V1) are at the same potential.

In other words the junction is a "virtual earth" summing point. Because of this virtual earth node the resistors, R_f and R_2 form a simple potential divider network across the non-inverting amplifier with the voltage gain of the circuit being determined by the ratios of R_2 and R_f as shown below.

Equivalent Potential Divider Network

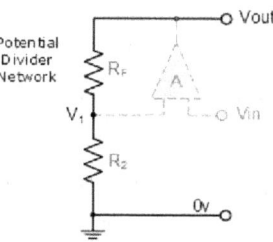

Then using the formula to calculate the output voltage of a potential divider network, we can calculate the closed-loop voltage gain (A_V) of the **Non-inverting Amplifier** as follows:

$$\text{Gain } (A_V) = V_{OUT} / V_{IN} = 1 + (R_F / R_2)$$

Demonstration:

$V_{IN} = V_1 = R_2 \times I_F$

$V_{OUT} = (R_2 + R_F) \times I_F$

$$\text{Gain } (A_V) = V_{OUT} / V_{IN} = (R_2 + R_F) \times I_F / R_2 \times I_F = (R_2 + R_F) / R_2 = 1 + R_F / R_2$$

Then the closed loop voltage gain of a **Non-inverting Operational Amplifier** will be given as:

$$A_{(v)} = 1 + \frac{R_F}{R_2}$$

We can see from the equation above, that the overall closed-loop gain of a non-inverting amplifier will always be greater but never less than one (unity), it is positive in nature and is determined by the ratio of the values of Rf and R2.

If the value of the feedback resistor Rf is zero, the gain of the amplifier will be exactly equal to one (unity). If resistor R2 is zero the gain will approach infinity, but in practice it will be limited to the operational amplifiers open-loop differential gain, (Ao).

5- Voltage Follower (Unity Gain Buffer)

If we made the feedback resistor, Rf equal to zero, ($Rf = 0$), and resistor R2 equal to infinity, (R2 = ∞), then the circuit would have a fixed gain of "1" as all the output voltage would be present on the inverting input terminal (negative feedback). This would then produce a special type of the non-inverting amplifier circuit called a **Voltage Follower** or also called a **"unity gain buffer"**.

As the input signal is connected directly to the non-inverting input of the amplifier the output signal is not inverted resulting in the output voltage being equal to the input voltage,

Vout = Vin. This then makes the **voltage follower** circuit ideal as a **Unity Gain Buffer** circuit because of its isolation properties.

The advantage of the unity gain voltage follower is that it can be used when impedance matching or circuit isolation is more important than amplification as it maintains the signal voltage. The input impedance of the voltage follower circuit is very high, typically above 1MΩ as it is equal to that of the operational amplifiers input resistance times its gain (Rin x Ao). Also its output impedance is very low since an ideal op-amp condition is assumed.

Non-inverting Voltage Follower

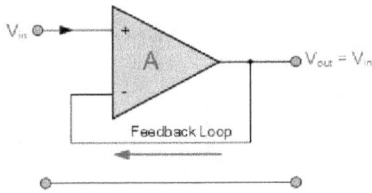

In this non-inverting circuit configuration, the input impedance Rin has increased to infinity and the feedback impedance Rf reduced to zero. The output is connected directly back to the negative inverting input so the feedback is 100% and Vin is exactly equal to Vout giving it a fixed gain of 1 or unity. As the input voltage Vin is applied to the non-inverting input the gain of the amplifier is given as:

$$V_{out} = A(V_{in})$$

$$(V_{in} = V+) \text{ and } (V_{out} = V-)$$

$$\text{therefore Gain, } (A_V) = \frac{V_{out}}{V_{in}} = +1$$

Since no current flows into the non-inverting input terminal the input impedance is infinite (ideal op-amp) and also no current flows through the feedback loop so any value of resistance may be placed in the feedback loop without affecting the characteristics of the circuit as no voltage is dissipated across it, zero current flows, zero voltage drop, zero power loss.

6- The Summing Amplifier

The Summing Amplifier is a very flexible circuit based upon the standard *Inverting Operational Amplifier* configuration. As its name suggests, the "summing amplifier" can be used for combining the voltage present on multiple inputs into a single output voltage.

We saw previously in the Inverting Operational Amplifier that the inverting amplifier has a single input voltage, (Vin) applied to the inverting input terminal. If we add more input resistors to the input, each equal in value to the original input resistor, Rin we end up with another operational amplifier circuit called a **Summing Amplifier**, "*summing inverter*" or even a "*voltage adder*" circuit as shown below.

Summing Amplifier Circuit

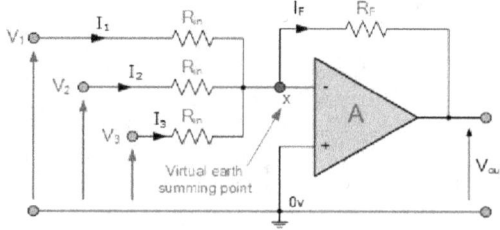

The output voltage, (Vout) now becomes proportional to the sum of the input voltages, V1, V2, V3 etc. Then we can modify the original equation for the inverting amplifier to take account of these new inputs thus:

$$I_F = I_1 + I_2 + I_3 = -\left[\frac{V1}{Rin} + \frac{V2}{Rin} + \frac{V3}{Rin}\right]$$

Inverting Equation: $Vout = -\frac{Rf}{Rin} \times Vin$

then, $-Vout = \left[\frac{R_F}{Rin}V1 + \frac{R_F}{Rin}V2 + \frac{R_F}{Rin}V3\right]$

However, if all the input impedances, (Rin) are equal in value, we can simplify the above equation to give an output voltage of:

Summing Amplifier Equation

$$-Vout = \frac{R_F}{R_{IN}}\left(V1 + V2 + V3....etc\right)$$

We now have an operational amplifier circuit that will amplify each individual input voltage and produce an output voltage signal that is proportional to the algebraic "SUM" of the three individual input voltages V_1, V_2 and V_3. We can also add more inputs if required as each individual input "see's" their respective resistance, Rin as the only input impedance.

This is because the input signals are effectively isolated from each other by the "virtual earth" node at the inverting input of the op-amp. A direct voltage addition can also be obtained when all the resistances are of equal value and Rf is equal to Rin.

A **Scaling Summing Amplifier** can be made if the individual input resistors are "NOT" equal. Then the equation would have to be modified to:

$$-Vout = V1\left(\frac{Rf}{R1}\right) + V2\left(\frac{Rf}{R2}\right) + V3\left(\frac{Rf}{R3}\right) ....etc$$

EEL 2003   Electrical / Electronics Department   LO # 2

To make the math's a little easier, we can rearrange the above formula to make the feedback resistor R_F the subject of the equation giving the output voltage as:

$$-Vout = Rf\left(\frac{V1}{R1} + \frac{V2}{R2} + \frac{V3}{R3}\right) \ldots etc$$

This allows the output voltage to be easily calculated if more input resistors are connected to the amplifiers inverting input terminal. The input impedance of each individual channel is the value of their respective input resistors, ie, R_1, R_2, R_3 ... etc.

Sometimes we need a summing circuit to just add together two or more voltage signals without any amplification. By putting all of the resistances of the circuit above to the same value R, the op-amp will have a voltage gain of unity and an output voltage equal to the direct sum of all the input voltages as shown:

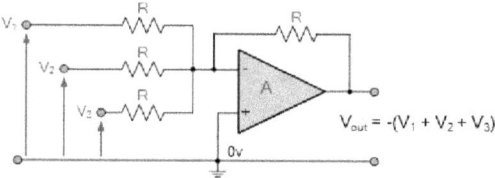

The **Summing Amplifier** is a very flexible circuit indeed, enabling us to effectively "Add" or "Sum" (hence its name) together several individual input signals. If the inputs resistors, R_1, R_2, R_3 etc, are all equal a "unity gain inverting adder" will be made. However, if the input resistors are of different values a "scaling summing amplifier" is produced which will output a weighted sum of the input signals.

Summing Amplifier Example

Find the output voltage of the following *Summing Amplifier* circuit.

Summing Amplifier

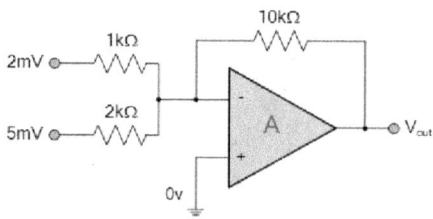

EEL 2003   Electrical / Electronics Department   LO # 2

Using the previously found formula for the gain of the circuit

$$\text{Gain}(A_v) = \frac{V_{out}}{V_{in}} = -\frac{R_f}{R_{in}}$$

We can now substitute the values of the resistors in the circuit as follows,

$$A_1 = \frac{10k\Omega}{1k\Omega} = -10$$

$$A_2 = \frac{10k\Omega}{2k\Omega} = -5$$

We know that the output voltage is the sum of the two amplified input signals and is calculated as:

$$V_{out} = (A_1 \times V_1) + (A_2 \times V_2)$$

$$V_{out} = (-10(2mV)) + (-5(5mV)) = -45mV$$

Then the output voltage of the **Summing Amplifier** circuit above is given as **-45 mV** and is negative as its an inverting amplifier.

Operational Amplifiers Summary

We know now that an **Operational amplifiers** is a very high gain DC differential amplifier that uses one or more external feedback networks to control its response and characteristics. We can connect external resistors or capacitors to the op-amp in a number of different ways to form basic "building Block" circuits such as, Inverting, Non-Inverting, Voltage Follower, Summing, Differential, Integrator and Differentiator type amplifiers.

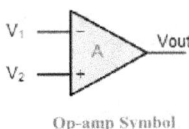

Op-amp Symbol

An "ideal" or perfect Operational Amplifier is a device with certain special characteristics such as:

- **infinite open-loop gain Ao,**
- **infinite input resistance Rin,**
- **zero output resistance Rout,**
- **infinite bandwidth 0 to ∞**
- **and zero offset (the output is exactly zero when the input is zero).**

There are a very large number of operational amplifier IC's available to suit every possible application from standard bipolar, precision, high-speed, low-noise, high-voltage, etc, in either standard configuration or with internal Junction FET transistors.

Operational amplifiers are available in IC packages of either single, dual or quad op-amps within one single device. The most commonly available and used of all operational amplifiers in basic electronic kits and projects is the industry standard µA-741.

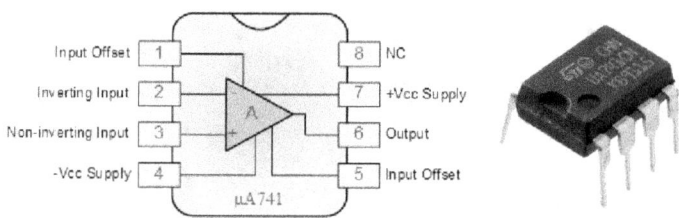

Typical Application

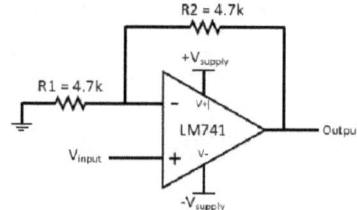

Figure 1. LM741 Noninverting Amplifier Circuit

Functional Block Diagram of LM 741

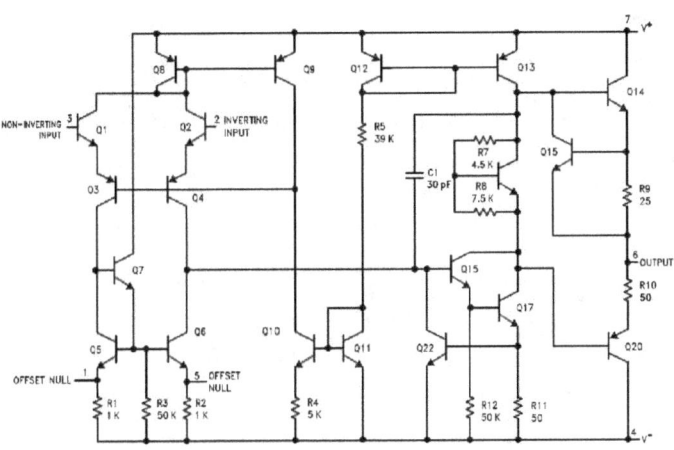

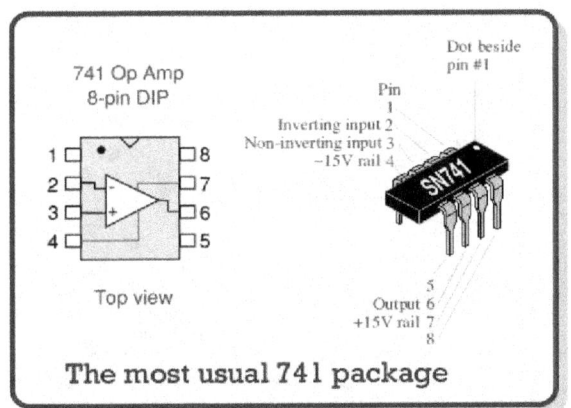

EEL 2003   Electrical Engineering Fundamentals   SO1/2

The Op-amp Comparator

1- Introduction

The comparator is an electronic decision making circuit that makes use of an operational amplifiers very high gain in its open-loop state, that is, there is no feedback resistor. The Op-amp comparator compares one analogue voltage level with another analogue voltage level, or some preset reference voltage, V_{REF} and produces an output signal based on this voltage comparison. In other words, the op-amp voltage comparator compares the magnitudes of two voltage inputs and determines which is the largest of the two.

$V_{OUT} = A_O (V+ - V-)$ where V+ and V- correspond to the voltages at the non-inverting and the inverting terminals respectively. The open loop gain is $A_o = 1 + (Rf / Rin)$

Due to this high open loop gain ($Rf = \infty$), then, the output from the comparator swings either fully to its positive supply rail, +Vcc or fully to its negative supply rail, -Vcc on the application of varying input signal which passes some preset threshold value.

The open-loop op-amp comparator is an analogue circuit that operates in its non-linear region as changes in the two analogue inputs, V+ and V- causes it to behave like a digital *bistable* device as triggering causes it to have two possible output states, +Vcc or -Vcc.

Consider the basic **op-amp voltage comparator** circuit below.

2- Op-amp Comparator Circuit

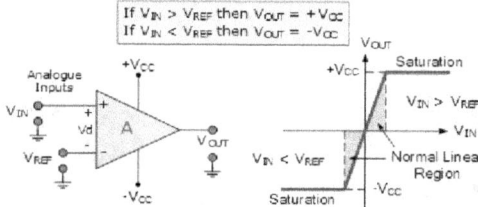

With reference to the op-amp comparator circuit above, lets first assume that V_{IN} is less than the DC voltage level at V_{REF}, ($V_{IN} < V_{REF}$). As the non-inverting (positive) input of the comparator is less than the inverting (negative) input, the output will be LOW and at the negative supply voltage, -Vcc resulting in a negative saturation of the output.

If we now increase the input voltage, V_{IN} so that its value is greater than the reference voltage V_{REF} on the inverting input, the output voltage rapidly switches HIGH towards the positive

a- Positive Voltage Comparator

The basic configuration for the positive voltage comparator, also known as a non-inverting comparator circuit detects when the input signal, V_{IN} is ABOVE or more positive than the reference voltage, V_{REF} producing an output at V_{OUT} which is HIGH as shown.

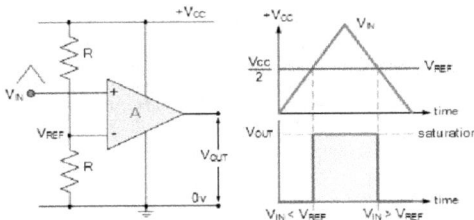

Non-inverting Comparator Circuit

In this non-inverting configuration, the reference voltage is connected to the inverting input of the operational amplifier with the input signal connected to the non-inverting input. To keep things simple, we have assumed that the two resistors forming the potential divider network are equal and: $R1 = R2 = R$. This will produce a fixed reference voltage which is one half that of the supply voltage, that is $Vcc/2$, while the input voltage is variable from zero to the supply voltage.

When V_{IN} is greater than V_{REF}, the op-amp comparators output will saturate towards the positive supply rail, Vcc. When V_{IN} is less than V_{REF} the op-amp comparators output will change state and saturate at the negative supply rail, 0v as shown.

b- Negative Voltage Comparator

The basic configuration for the negative voltage comparator, also known as an inverting comparator circuit detects when the input signal, V_{IN} is BELOW or more negative than the reference voltage, V_{REF} producing an output at V_{OUT} which is HIGH as shown.

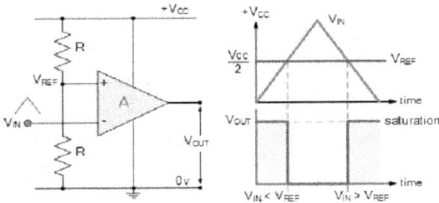

Inverting Comparator Circuit

In the inverting configuration, which is the opposite of the positive configuration above, the reference voltage is connected to the non-inverting input of the operational amplifier while the input signal is connected to the inverting input. Then when V_{IN} is less than V_{REF} the op-amp comparators output will saturate towards the positive supply rail, Vcc.

Likewise the reverse is true, when V_{IN} is greater than V_{REF}, the op-amp comparators output will change state and saturate towards the negative supply rail, 0v.

Then depending upon which op-amp inputs we use for the signal and the reference voltage, we can produce an inverting or non-inverting output. We can take this idea of detecting either a negative or positive going signal one step further by combining the two op-amp comparator circuits above to produce a window comparator circuit.

EEL 2003   Electrical Engineering Fundamentals   SO1/2

PWM Signal Generators

1. Introduction

PWM, or Pulse Width Modulation, is a method of controlling the amount of power to a load without having to dissipate any power in the load driver.

Imagine a 10W light bulb load supplied from a battery. In this case the battery supplies 10W of power, and the light bulb converts this 10W into light and heat. No power is lost anywhere else in the circuit. If we wanted to dim the light bulb, so it only absorbed 5W of power, we could place a resistor in series which absorbed 5W, then the light bulb could absorb the other 5W. This would work, but the power dissipated in the resistor not only makes it get very hot, but is wasted. The battery is still supplying 10W.

An alternative way is to switch the light bulb on and off very quickly so that it is only on for half of the time. Then the *average* power taken by the light bulb is still only 5W, and the average power supplied by the battery is only supplying 5W also. If we wanted the bulb to take 6W, we could leave the switch on for a little longer than the time it was off, then a little more average power will be delivered to the bulb.

This on-off switching is called PWM. The amount of power delivered to the load is proportional to the percentage of time that the load is switched on.

2- PWM Generation

A block diagram of an analogue PWM generator is shown below:

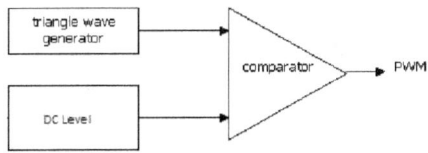

We are starting at the output because this is the easy bit. The diagram below shows how comparing a ramping waveform with a DC level produces the PWM waveform that we require. The lower the DC level is, the wider the PWM pulses are. The DC level is the reference level.

The DC signal can range between the minimum and maximum voltages of the triangle wave.

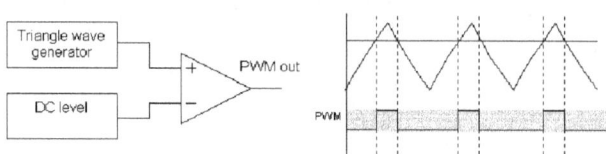

When the triangle waveform voltage is greater than the DC level, the output of the op-amp swings high, and when it is lower, the output swings low.

3- Pulse Width Modulated Waveform

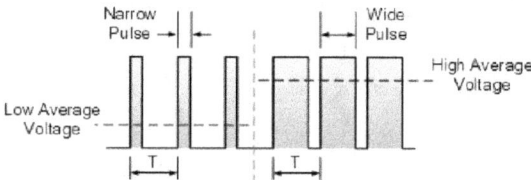

The power applied to the load (motor, bulb ...) can be controlled by varying the width of these applied pulses and thereby varying the average DC voltage applied to the load terminals. By changing or modulating the timing of these pulses the speed of the motor can be controlled, ie, the longer the pulse is "ON", the faster the motor will rotate and likewise, the shorter the pulse is "ON" the slower the motor will rotate.

4- Duty cycle:

The **duty cycle** is **defined** as the ratio between the pulse duration ( ) and the period ( ) of a rectangular waveform

$$\text{Duty Cycle \%} = 100 \times (\text{Time ON}) / (\text{Time ON} + \text{Time OFF}) = 100 \times \text{Time ON} / T$$

Electrical motors typically use less than a 100% duty cycle. For example, if a motor runs for one out of 100 seconds, or 1/100 of the time, then, its duty cycle is 1/100, or 1 percent

If the motor runs for 20 seconds out of 100, the duty cycle will be:

$$\text{Duty cycle \%} = (20/100) \times 100 = 20\%$$

https://www.youtube.com/watch?v=Lf7JJAAZxEU

EEL 2003 Electrical Engineering Fundamentals SO1/2

5- Generator Pulse

Generate pulses for carrier-based two-level pulse width modulator (PWM) in converter bridge

The PWM Generator block generates pulses for carrier-based pulse width modulation (PWM) converters using two-level topology. The block can be used to fire the forced-commutated devices (FETs, GTOs, or IGBTs) of single-phase, two-phase, three-phase, two-level bridges or a combination of two three-phase bridges.

The amplitude (modulation), phase, and frequency of the reference signals are set to control the output voltage (on the AC terminals) of the bridge connected to the PWM Generator block.

The two pulses firing the two devices of a given arm bridge are complementary. For example, pulse 4 is low (0) when pulse 3 is high (1). This is illustrated in the next two figures.

The following figure displays the two pulses generated by the PWM Generator block when it is programmed to control a one-arm bridge

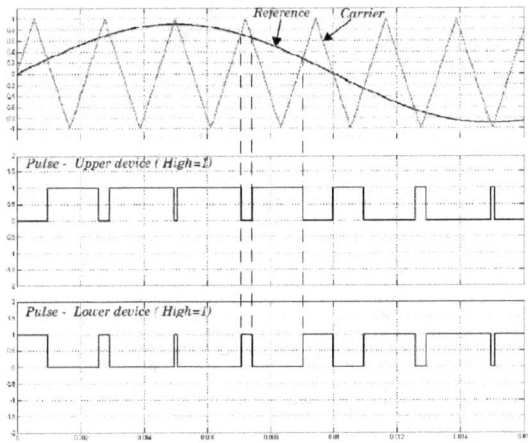

The triangular carrier signal is compared with the sinusoidal modulating signal. When the modulating signal is greater than the carrier pulse 1 is high (1) and pulse 2 is low (0).

EEL 2003   Electrical Engineering Fundamentals   SO1/2

The 555 Timer

1- Introduction

There are also dedicated IC's especially designed to accurately produce the required output waveform with the addition of just a few extra timing components. One such device that has been around since the early days of IC's and has itself become something of an industry "standard" is the 555 Timer Oscillator which is more commonly called the "555 Timer".

The 555 timer chip is extremely robust and stable 8-pin device that can be operated either as a very accurate **Monostable**, **Bistable** or **Astable** Multivibrator to produce a variety of applications such as one-shot or delay timers, pulse generation, LED and lamp flashers, alarms and tone generation, logic clocks, frequency division, power supplies and converters etc, in fact any circuit that requires some form of time control as the list is endless.

2- Monostable 555 Timer

A simplified "block diagram" representing the internal circuitry of the **555 timer** is given below with a brief explanation of each of its connecting pins to help provide a clearer understanding of how it works.

The **Monostable 555 Timer** circuit triggers on a negative-going pulse applied to pin 2 and this trigger pulse must be much shorter than the output pulse width allowing time for the timing capacitor to charge and then discharge fully. Once triggered, the 555 Monostable will remain in this "HIGH" unstable output state until the time period set up by the $R_1 \times C_1$ network has elapsed. The amount of time that the output voltage remains "HIGH" or at a logic "1" level, is given by the following time constant equation.

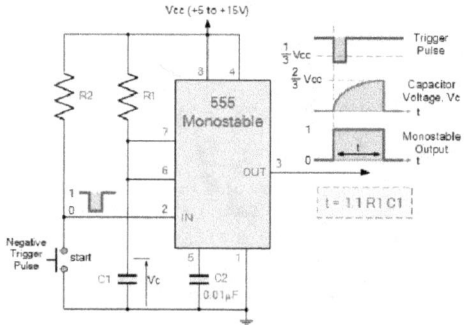

The amount of time that the output voltage remains "HIGH" or at a logic "1" level, is given by the following time constant equation.

$$\tau = 1.1 R_1 C_1$$

Where, t is in seconds, R is in Ω's and C in Farads.

555 Timer Example No1

A **Monostable 555 Timer** is required to produce a time delay within a circuit. If a 10uF timing capacitor is used, calculate the value of the resistor required to produce a minimum output time delay of 500ms.

500ms is the same as saying 0.5s so by rearranging the formula above, we get the calculated value for the resistor, R as:

$$R = \frac{t}{1.1C} = \frac{0.5}{1.1 \times 10uF} = \frac{0.5}{1.1 \times 10 \times 10^{-6}} = 45.5k\Omega$$

The calculated value for the timing resistor required to produce the required time constant of 500ms is therefore, 45.5KΩ's. However, the resistor value of 45.5KΩ's does not exist as a standard value resistor, so we would need to select the nearest preferred value resistor of 47kΩ's which is available in all the standard ranges of tolerance from the E12 (10%) to the E96 (1%), giving us a new recalculated time delay of 517ms.

3- Astable 555 Timer

Whereas the 555 monostable circuit stopped after a preset time waiting for the next trigger pulse to start over again, in order to get the 555 Oscillator to operate as an astable multivibrator it is necessary to continuously re-trigger the 555 IC after each and every timing cycle.

This re-triggering is basically achieved by connecting the *trigger* input (pin 2) and the *threshold* input (pin 6) together, thereby allowing the device to act as an astable oscillator. Then the 555 Oscillator has no stable states as it continuously switches from one state to the other. Also the single timing resistor of the previous monostable multivibrator circuit has been split into two separate resistors, R1 and R2 with their junction connected to the *discharge* input (pin 7) as shown below.

a/ 555 Timer Operation

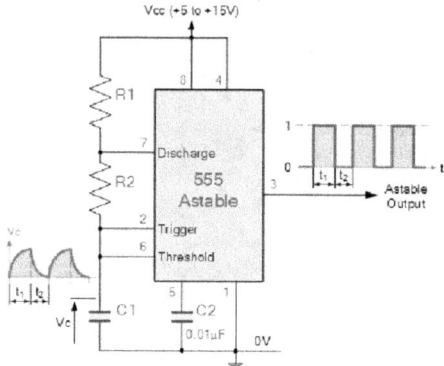

Basic Astable 555 Oscillator Circuit

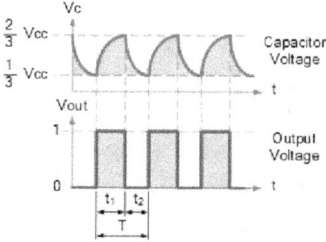

In the **555 Oscillator** circuit above, pin 2 and pin 6 are connected together allowing the circuit to re-trigger itself on each and every cycle allowing it to operate as a free running oscillator. During each cycle capacitor, C charges up through both timing resistors, R1 and R2 but discharges itself only through resistor, R2 as the other side of R2 is connected to the *discharge* terminal, pin 7.

b/ Astable 555 Oscillator Charge and Discharge Times

$$t_1 = 0.693(R_1 + R_2).C$$
and
$$t_2 = 0.693 \times R_2 \times C$$

Where, R is in Ω's and C in Farads.

When connected as an astable multivibrator, the output from the **555 Oscillator** will continue indefinitely charging and discharging between 2/3Vcc and 1/3Vcc until the power supply is removed. As with the monostable multivibrator these charge and discharge times and therefore the frequency are independent on the supply voltage.

The duration of one full timing cycle is therefore equal to the sum of the two individual times that the capacitor charges and discharges added together and is given as:

c/ 555 Oscillator Cycle Time

$$T = t_1 + t_2 = 0.693(R_1 + 2R_2).C$$

The output frequency of oscillations can be found by inverting the equation above for the total cycle time giving a final equation for the output frequency of an Astable 555 Oscillator as:

d/ 555 Oscillator Frequency Equation

$$f = \frac{1}{T} = \frac{1.44}{(R_1 + 2R_2).C}$$

By altering the time constant of just one of the RC combinations, the **Duty Cycle** better known as the "Mark-to-Space" ratio of the output waveform can be accurately set and is given as the ratio of resistor R2 to resistor R1. The Duty Cycle for the 555 Oscillator, which is the ratio of the "ON" time divided by the "OFF" time is given by:

e/ 555 Oscillator Duty Cycle

$$\text{Duty Cycle} = \frac{T_{ON}}{T_{OFF} + T_{ON}} = \frac{R_1 + R_2}{(R_1 + 2R_2)} \%$$

The duty cycle has no units as it is a ratio but can be expressed as a percentage (%). If both timing resistors, R1 and R2 are equal in value, then the output duty cycle will be 2:1 that is, 66% ON time and 33% OFF time with respect to the period.

f- 555 Oscillator Example No1

An **Astable 555 Oscillator** is constructed using the following components, R1 = 1kΩ, R2 = 2kΩ and capacitor C = 10uF. Calculate the output frequency from the 555 oscillator and the duty cycle of the output waveform.

t_1 – capacitor charge "ON" time is calculated as:

$$t_1 = 0.693(R_1 + R_2).C$$
$$= 0.693(1000 + 2000) \times 10 \times 10^{-6}$$
$$= 0.021s = 21ms$$

t_2 – capacitor discharge "OFF" time is calculated as:

$$t_2 = 0.693\, R_2.C$$
$$= 0.693 \times 2000 \times 10 \times 10^{-6}$$
$$= 0.014s = 14ms$$

Total periodic time (T) is therefore calculated as:

$$T = t_1 + t_2 = 21ms + 14ms = 35ms$$

The output frequency, f is therefore given as:

$$f = \frac{1}{T} = \frac{1}{35ms} = 28.6Hz$$

Giving a duty cycle value of:

$$\text{Duty Cycle} = \frac{R_1 + R_2}{(R_1 + 2R_2)} = \frac{1000 + 2000}{(1000 + 2 \times 2000)} = 0.6 \text{ or } 60\%$$

EEL 2003   Electrical Engineering Fundamentals   SO1/2

4- 555 Timer Transistor Driver

We said earlier that the maximum output current to either sink or source the load current via pin 3 is about 200mA at the maximum supply voltage, and this value is more than enough to drive or switch other logic IC's, LED's or small lamps, etc. But what if we wanted to switch or control higher power devices such as motors, electromagnets, relays or loudspeakers. Then we would need to use a **Transistor** to amplify the 555 timers output in order to provide a sufficiently high enough power to drive the load.

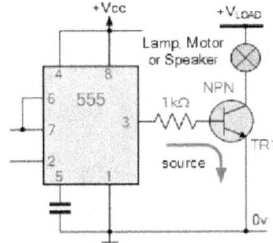

The transistor in the two examples above, can be replaced with a Power MOSFET device or Darlington transistor if the load current is high. When using an inductive load such as a motor, relay or electromagnet, it is advisable to connect a freewheel (or flywheel) diode directly across the load terminals to absorb any back emf voltages generated by the inductive device when it changes state.

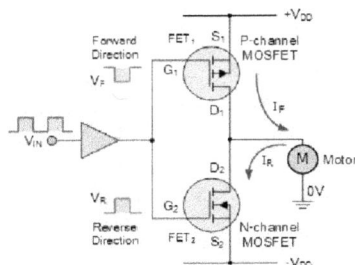

Complementary MOSFET Motor Controller

The two MOSFETs are configured to produce a bi-directional switch from a dual supply with the motor connected between the common drain connection and ground reference. When the input is LOW the P-channel MOSFET is switched-ON as its gate-source junction is negatively biased so the motor rotates in one direction. Only the positive $+V_{DD}$ supply rail is used to drive the motor.

When the input is HIGH, the P-channel device switches-OFF and the N-channel device switches-ON as its gate-source junction is positively biased. The motor now rotates in the opposite direction because the motors terminal voltage has been reversed as it is now supplied by the negative $-V_{DD}$ supply rail.

Complementary MOSFET Motor Control Table

MOSFET 1	MOSFET 2	Motor Function
OFF	OFF	Motor Stopped (OFF)
ON	OFF	Motor Rotates Forward
OFF	ON	Motor Rotates Reverse
ON	ON	NOT ALLOWED

EEL 2003   Electrical Engineering Fundamentals   SO1/2

PWM Motor Control

The magnetic field produced by the stator's permanent magnets is fixed and therefore can not be changed but if we change the strength of the armatures electromagnetic field by controlling the current flowing through the windings more or less magnetic flux will be produced resulting in a stronger or weaker interaction and therefore a faster or slower speed.

Then the rotational speed of a DC motor (N) is proportional to the back emf (V_b) of the motor divided by the magnetic flux (which for a permanent magnet is a constant) times an electromechanical constant depending upon the nature of the armatures windings (K_e) giving us the equation of: $N \propto V/K_e\phi$.

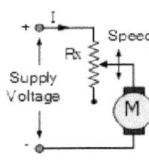 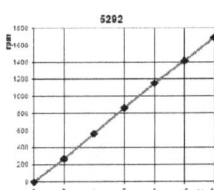

So how do we control the flow of current through the motor. Well many people attempt to control the speed of a DC motor using a large variable resistor (Rheostat) in series with the motor as shown.

, it generates a lot of heat and wasted power in the resistance. One simple and easy way to control the speed of a motor is to regulate the amount of voltage across its terminals and this can be achieved using "**Pulse Width Modulation**" or PWM.

As its name suggests, pulse width modulation speed control works by driving the motor with a series of "ON-OFF" pulses and varying the duty cycle, the fraction of time that the output voltage is "ON" compared to when it is "OFF", of the pulses while keeping the frequency constant.

The power applied to the motor can be controlled by varying the width of these applied pulses and thereby varying the average DC voltage applied to the motors terminals. By changing or modulating the timing of these pulses the speed of the motor can be controlled, ie, the longer the pulse is "ON", the faster the motor will rotate and likewise, the shorter the pulse is "ON" the slower the motor will rotate.

In other words, the wider the pulse width, the more average voltage applied to the motor terminals, the stronger the magnetic flux inside the armature windings and the faster the motor will rotate and this is shown below.

Summary Operational Amplifier

Inverting Op-amp

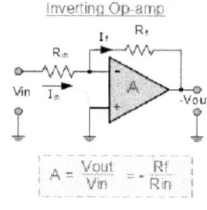

$$A = \frac{V_{out}}{V_{in}} = -\frac{R_f}{R_{in}}$$

Non-inverting Op-amp

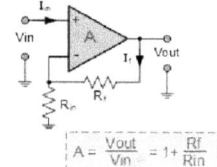

$$A = \frac{V_{out}}{V_{in}} = 1 + \frac{R_f}{R_{in}}$$

Summing Op-amp

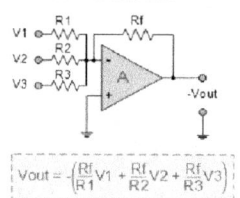

$$V_{out} = -\left(\frac{R_f}{R_1}V_1 + \frac{R_f}{R_2}V_2 + \frac{R_f}{R_3}V_3\right)$$

Non-inverting Voltage Follower

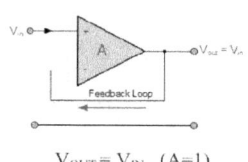

$$V_{OUT} = V_{IN} \quad (A=1)$$

Op-amp Comparator Circuit

If $V_{IN} > V_{REF}$ then $V_{OUT} = +V_{CC}$
If $V_{IN} < V_{REF}$ then $V_{OUT} = -V_{CC}$

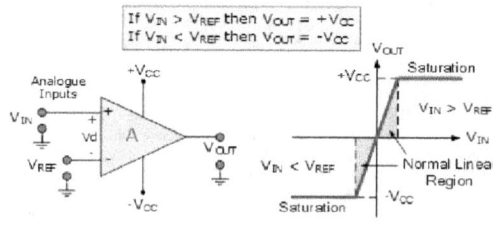

C. Part 3: Analyse the use of uncontrolled and controlled single-phase rectifiers in various electrical power applications.

1. Determine the performance characteristics of uncontrolled single-phase, half-wave and full-wave rectifiers, with resistive and inductive loads.
2. Determine the performance characteristics of controlled single-phase, half-wave and full-wave rectifiers with resistive and inductive loads.
3. Determine the change in power factor when using uncontrolled and controlled rectifiers. Define input distortion and displacement factor.
4. Describe how power inversion may be achieved by varying the firing angle in controlled rectifiers.

DIODE CIRCUITS OR UNCONTROLLED RECTIFIER

2.1 Introduction

Because of their ability to conduct current in one direction, diodes are used in rectifier circuits. The definition of rectification process is " *the process of converting the alternating voltages and currents to direct currents and the device is known as rectifier*" It is extensively used in charging batteries; supply DC motors, electrochemical processes and power supply sections of industrial components.

The most famous diode rectifiers have been analyzed in the following sections. Circuits and waveforms drawn with the help of PSIM simulation program.

There are two different types of uncontrolled rectifiers or diode rectifiers, half wave and full wave rectifiers. Full-wave rectifiers has better performance than half wave rectifiers. But the main advantage of half wave rectifier is its need to less number of diodes than full wave rectifiers. The main disadvantages of half wave rectifier are:

1- High ripple factor.
2- Low rectification efficiency.
3- Low transformer utilization factor, and.
4- DC saturation of transformer secondary winding.

2.2 Performance Parameters

In most rectifier applications, the power input is sine-wave voltage provided by the electric utility that is converted to a DC voltage and AC components. The AC components are undesirable and must be kept away from the load. Filter circuits or any other harmonic reduction technique should be installed between the electric utility and the rectifier and between the rectifier output and the load that filters out the undesired component and allows useful components to go through. So, careful analysis has to be done before building the rectifier. The analysis requires define the following terms:

The average value of the output voltage, V_{dc}.
The average value of the output current, I_{dc}.
The *rms* value of the output voltage, V_{rms}.
The *rms* value of the output current, I_{rms}

The output DC power, $P_{dc} = V_{dc} * I_{dc}$ (2.1)
The output AC power, $P_{ac} = V_{rms} * I_{rms}$ (2.2)
The effeciency or rectification ratio is defiend as $\eta = P_{dc} / P_{ac}$ (2.3)

The output voltage can be considered as being composed of two components (1) the DC component and (2) the AC component or ripple. The effective (*rms*) value of the AC component of output voltage is defined as:-

$$V_{ac} = \sqrt{V_{rms}^2 - V_{dc}^2} \qquad (2.4)$$

Single-Phase Half Wave Diode Rectifier With Resistive Load

Fig.2.1 shows a single-phase half-wave diode rectifier with pure resistive load. Assuming sinusoidal voltage source, V_S the diode begins to conduct when its anode voltage is greater than its cathode voltage as a result, the load current flows. So, the diode will be in "ON" state in positive voltage half cycle and in "OFF" state in negative voltage half cycle. Fig.2.2 shows various current and voltage waveforms of half wave diode rectifier with resistive load. These waveforms show that both the load voltage and current have high ripples. For this reason, single-phase half-wave diode rectifier has little practical significance.

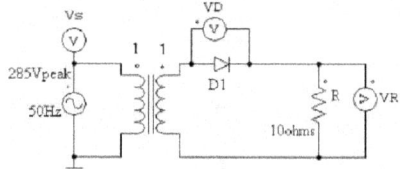

Fig.2.1 Single-phase half-wave diode rectifier with resistive load.

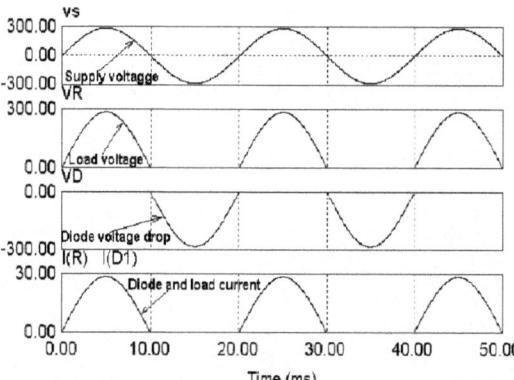

Fig.2.2 Various waveforms for half wave diode rectifier with resistive load.

The average or DC output voltage can be obtained by considering the waveforms shown in Fig.2.2 as following: $V_{dc} = \frac{1}{2\pi} \int_0^\pi V_m \sin \omega t \, d\omega t = \frac{V_m}{\pi}$ (2.12)

Where, V_m is the maximum value of supply voltage.

Because the load is resistor, the average or DC component of load current is: $I_{dc} = \frac{V_{dc}}{R} = \frac{V_m}{\pi R}$

The root mean square (rms) value of a load voltage is defined as:

$V_{rms} = \sqrt{\frac{1}{2\pi} \int_0^\pi V_m^2 \sin^2 \omega t \, d\omega t} = \frac{V_m}{2}$ (2.14)

Similarly, the root mean square (rms) value of a load current is defined as: $I_{rms} = \frac{V_{rms}}{R} = \frac{V_m}{2R}$

It is clear that the rms value of the transformer secondary current, I_S is the same as that of the load and diode currents

Then $I_S = I_D = \frac{V_m}{2R}$ (2.15)

Where, I_D is the rms value of diode current.

Example 1: The rectifier shown in Fig.2.1 has a pure resistive load of R. Determine (a) The efficiency, (b) Form factor (c) Ripple factor (d) Peak inverse voltage (PIV) of diode D1.
Solution: From Fig.2.2, the average output voltage V_{dc} is defiend as:

$V_{dc} = \frac{1}{2\pi} \int_0^\pi V_m \sin(\omega t) \, d\omega t = \frac{V_m}{2\pi}(-\cos\pi - \cos(0)) = \frac{V_m}{\pi}$ Then, $I_{dc} = \frac{V_{dc}}{R} = \frac{V_m}{\pi R}$

$V_{rms} = \sqrt{\frac{1}{2\pi} \int_0^\pi (V_m \sin \omega t)^2} = \frac{V_m}{2}$, $I_{rms} = \frac{V_m}{2R}$ and, $V_S = \frac{V_m}{\sqrt{2}}$

The rms value of the transformer secondery current is the same as that of the load: $I_S = V_m/2R$
Then, the efficiency or rectification ratio is:

$\eta = \frac{P_{dc}}{P_{ac}} = \frac{V_{dc} * I_{dc}}{V_{rms} * I_{rms}} = \frac{\frac{V_m}{\pi} * \frac{V_m}{\pi R}}{\frac{V_m}{2} * \frac{V_m}{2R}} = 40.53\%$

2.4 Single-Phase Full-Wave Diode Rectifier

The full wave diode rectifier can be designed with a center-taped transformer as shown in Fig.2.8, where each half of the transformer with its associated diode acts as half wave rectifier or as a bridge diode rectifier as shown in Fig. 2.12. The advantage and disadvantage of center-tap diode rectifier is shown below:

ADVANTAGES
- The need for center-tapped transformer is eliminated.
- The output is twice that of the center tapped circuit for the same secondary voltage, and,
- The peak inverse voltage is one half of the center-tap circuit.

DISADVANTAGES
- It requires four diodes instead of two, in full wave circuit, and,
- There are always two diodes in series are conducting. Therefore, total voltage drop in the internal resistance of the diodes and losses are increased.

The following sections explain and analyze these rectifiers.

2.4.1 Center-Tap Diode Rectifier With Resistive Load

In the center tap full wave rectifier, current flows through the load in the same direction for both half cycles of input AC voltage. The circuit shown in Fig.2.8 has two diodes D1 and D2 and a center tapped transformer. The diode D1 is forward bias "ON" and diode D2 is reverse bias "OFF" in the positive half cycle of input voltage and current flows from point a to point b. Whereas in the negative half cycle the diode D1 is reverse bias "OFF" and diode D2 is forward bias "ON" and again current flows from point a to point b. Hence DC output is obtained across the load.

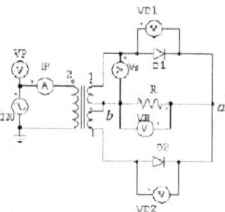

Fig.2.8 Center-tap diode rectifier

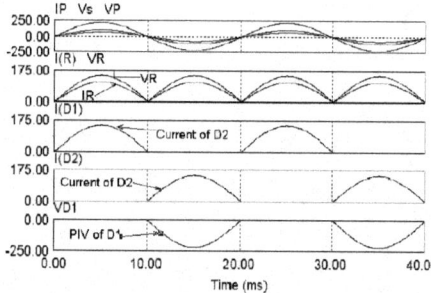

Fig.2.9 Various current and voltage waveforms for center-tap diode rectifier with resistive load.

In case of pure resistive load, Fig.2.9 shows various current and voltage waveform for converter in Fig.2.8. The average and *rms* output voltage and current can be obtained from the waveforms shown in Fig.2.9 as shown in the following:

$$V_{dc} = \frac{1}{\pi}\int_0^\pi V_m \sin \omega t\, d\omega t = \frac{2V_m}{\pi} \quad \text{and} \quad I_{dc} = \frac{2V_m}{\pi R} \tag{2.36}$$

$$V_{rms} = \sqrt{\frac{1}{\pi}\int_0^\pi (V_m \sin \omega t)^2\, d\omega t} = \frac{V_m}{\sqrt{2}} \quad \text{and} \quad I_{rms} = \frac{V_m}{\sqrt{2} R} \tag{2.38}$$

PIV of each diode $= 2V_m$ \hfill (2.40)

$$V_S = \frac{V_m}{\sqrt{2}} \tag{2.41}$$

The *rms* value of the transformer secondery current is the same as that of the diode:

$$I_S = I_D = \frac{V_m}{2R} \tag{2.41}$$

Example 3. The rectifier in Fig.2.8 has a purely resistive load of R. Determine (a) The efficiency, (b) Form factor (c) Ripple factor (d) Peak inverse voltage (PIV) of diode D1.
Solution:- The efficiency or rectification ratio is

$$\eta = \frac{P_{dc}}{P_{ac}} = \frac{V_{dc} * I_{dc}}{V_{rms} * I_{rms}} = \frac{\frac{2V_m}{\pi} * \frac{2V_m}{\pi R}}{\frac{V_m}{\sqrt{2}} * \frac{V_m}{\sqrt{2} R}} = 81.05\%$$

2.4.3 Single-Phase Full Bridge Diode Rectifier With Resistive Load

Another alternative in single-phase full wave rectifier is by using four diodes as shown in Fig.2.12 which known as a single-phase full bridge diode rectifier. It is easy to see the operation of these four diodes. The current flows through diodes D1 and D2 during the positive half cycle of input voltage (D3 and D4 are "OFF"). During the negative one, diodes D3 and D4 conduct (D1 and D2 are "OFF").

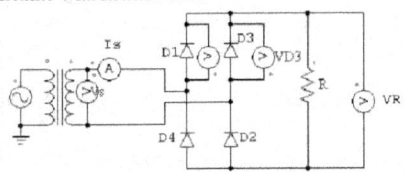

Fig.2.12 Single-phase full bridge diode rectifier.

In positive half cycle the supply voltage forces diodes D1 and D2 to be "ON". In same time it forces diodes D3 and D4 to be "OFF". So, the current moves from positive point of the supply voltage across D1 to the point a of the load then from point b to the negative marked point of the supply voltage through diode D2. In the negative voltage half cycle, the supply voltage forces the diodes D1 and D2 to be "OFF". In same time it forces diodes D3 and D4 to be "ON". So, the current moves from negative marked point of the supply voltage across D3 to the point a of the load then from point b to the positive marked point of the supply voltage through diode D4. So, it is clear that the load currents moves from point a to point b in both positive and negative half cycles of supply voltage. So, a DC output current can be obtained at the load in both positive and negative halves cycles of the supply voltage. The complete waveforms for this rectifier is shown in Fig.2.13.

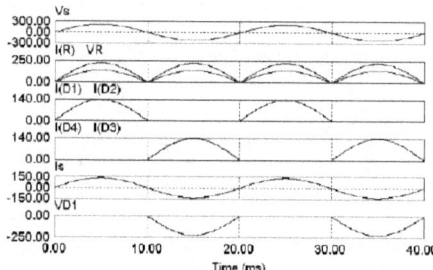

Fig.2.13 Various current and voltage waveforms of Full bridge single-phase diode rectifier.

Example 4 The rectifier shown in Fig.2.12 has a purely resistive load of $R=15\ \Omega$ and. $V_s=300\sin 314\ t$ and unity transformer ratio. Determine (a) The efficiency, (b) Form factor, (c) Ripple factor, (d) The peak inverse voltage, (PIV) of each diode. , and. (e) Input power factor.
Solution: $V_m = 300\ V$

$$V_{dc} = \frac{1}{\pi}\int_0^{\pi} V_m \sin \omega t\ d\omega t = \frac{2V_m}{\pi} = 190.956\ V, \quad I_{dc} = \frac{2V_m}{\pi R} = 12.7324\ A$$

$$V_{rms} = \left[\frac{1}{\pi}\int_0^{\pi}(V_m \sin \omega t)^2\ d\omega t\right]^{1/2} = \frac{V_m}{\sqrt{2}} = 212.132\ V, \quad I_{rms} = \frac{V_m}{\sqrt{2}R} = 14.142A$$

(a) $\eta = \dfrac{P_{dc}}{P_{ac}} = \dfrac{V_{dc}\ I_{dc}}{V_{rms}\ I_{rms}} = 81.06\ \%$

EEL 2003   Electrical / Electronics Department   SO1-2

THYRISTOR CONVERTERS OR CONTROLLED CONVERTERS
3.1 Introduction
The controlled rectifier circuit is divided into three main circuits:-
- **(1) Power Circuit**
  This is the circuit contains voltage source, load and switches as diodes, thyristors or IGBTs.
- **(2) Control Circuit**
  This circuit is the circuit, which contains the logic of the firing of switches that may contains amplifiers, logic gates and sensors.
- **(3) Triggering circuit**
  This circuit lies between the control circuit and power thyristors. Sometimes this circuit called *switch drivers circuit*. This circuit contains buffers, opt coupler or pulse transformers. The main purpose of this circuit is to *separate between the power circuit and control circuit*.

The method of switching off the thyristor is known as Thyristor commutation. The thyristor can be turned off by reducing its forward current below its holding current or by applying a reverse voltage across it. The commutation of thyristor is classified into two types:-

1- Natural Commutation
If the input voltage is AC, the thyristor current passes through a natural zero, and a reverse voltage appear across the thyristor, which in turn automatically turned off the device due to the natural behavior of AC voltage source. This is known as natural commutation or line commutation. This type of commutation is applied in AC voltage controller rectifiers and cycloconverters.

2- Forced Commutation
In DC thyristor circuits, if the input voltage is DC, the forward current of the thyristor is forced to zero by an additional circuit called commutation circuit to turn off the thyristor. This technique is called forced commutation. Normally this method for turning off the thyristor is applied in choppers.

There are many thyristor circuits we can not present all of them. In the following items we are going to present and analyze the most famous thyristor circuits.

3.2 Half Wave Single Phase Controlled Rectifier
3.2.1 Half Wave Single Phase Controlled Rectifier With Resistive Load
The circuit with single SCR is similar to the single diode circuit, the difference being that an SCR is used in place of the diode. Most of the power electronic applications operate at a relative high voltage and in such cases; the voltage drop across the SCR tends to be small. It is quite often justifiable to assume that the conduction drop across the SCR is zero when the circuit is analyzed. It is also justifiable to assume that the current through the SCR is zero when it is not conducting. It is known that the SCR can block conduction in either direction. The explanation and the analysis presented below are based on the ideal SCR model. All simulation carried out by using PSIM computer simulation program.

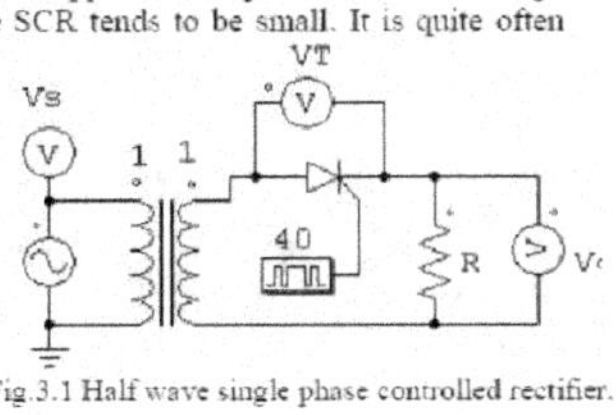

A circuit with a single SCR and resistive load is shown in Fig.3.1. The source v_s is an alternating sinusoidal

Fig.3.1 Half wave single phase controlled rectifier.

source. If $v_z = V_m \sin(\omega t)$, v_z is positive when $0 < \omega t < \pi$, and v_z is negative when $\pi < \omega t < 2\pi$. When v_s starts become positive, the SCR is forward-biased but remains in the blocking state till it is triggered. If the SCR is triggered at $\omega t = \alpha$, then α is called the *firing angle*. When the SCR is triggered in the forward-bias state, it starts conducting and the positive source keeps the SCR in conduction till ωt reaches π radians. At that instant, the current through the circuit is zero. After that the current tends to flow in the reverse direction and the SCR blocks conduction. The entire applied voltage now appears across the SCR. Various voltages and currents waveforms of the half-wave controlled rectifier with resistive load are shown in Fig.3.2 for $\alpha=40°$. FFT components for load voltage and current of half wave single phase controlled rectifier with resistive load at $\alpha=40°$ are shown in Fig.3.3. It is clear from Fig.3.3 that the supply current containes DC component and all other harmonic components which makes the supply current highly distorted. For this reason, this converter does not have acceptable practical applications.

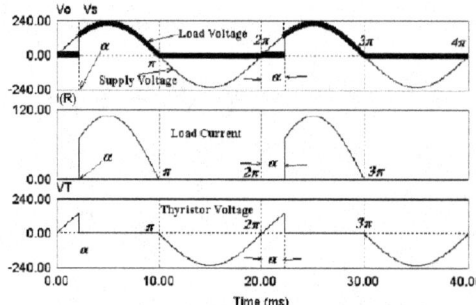

Fig.3.2 Various voltages and currents waveforms for half wave single-phase controlled rectifier with resistive load at $\alpha=40°$.

The average voltage, V_{dc}, across the resistive load can be obtained by considering the waveform shown in Fig.3.2. $\quad V_{dc} = \frac{1}{2\pi} \int_{\alpha}^{\pi} V_m \sin(\omega t) \, d\omega t = \frac{V_m}{2\pi}(-\cos\pi + \cos(\alpha)) = \frac{V_m}{2\pi}(1+\cos\alpha)$ \quad (3.1)

The maximum output voltage and can be acheaved when $\alpha = 0$ which is the same as diode case which obtained before in (2.12). $V_{dm} = V_m / \pi$ \hfill (3.2)
The normalized output voltage is the DC voltage devideded by maximum DC voltage, V_{dn} which can be obtained as shown in equation (3.3). $\quad V_n = V_{dc}/V_m = 0.5 (1+\cos\alpha)$ \hfill (3.3)

The *rms* value of the output voltage is shown in the following equation:-

$$V_{rms} = \sqrt{\frac{1}{2\pi}\int_{\alpha}^{\pi}(V_m \sin(\omega t))^2 \, d\omega t} = \frac{V_m}{2}\sqrt{\frac{1}{\pi}\left(\pi - \alpha + \frac{\sin(2\alpha)}{2}\right)} \quad (3.3)$$

The *rms* value of the transformer secondery current and load is: $I_s = V_{rms}/R$ \hfill (3.4)

Example 1 In the rectifier shown in Fig.3.1 it has a load of $R=15\ \Omega$ and, $V_s=220\ sin\ 314\ t$ and unity transformer ratio. If it is required to obtain an average output voltage of 70% of the maximum possible output voltage, calculate:- (a) The firing angle, α, (b) The efficiency, (c) Ripple factor (d) Peak inverse voltage (PIV) of the thyristor

Solution: (a) V_{dm} is the maximum output voltage and can be acheaved when $\alpha=0$. The normalized output voltage is shown in equation (3.3) which is required to be 70%. Then,

$$V_n = \frac{V_{dc}}{V_{dm}} = 0.5\,(1+\cos\alpha) = 0.7\ .\ \text{Then},\ \alpha = 66.42° = 1.15925\ rad.$$

(b) $V_m = 220\ V$. $V_{dc} = 0.7 * V_{dm} = 0.7 * \frac{V_m}{\pi} = 49.02\ V$. $I_{dc} = \frac{V_{dc}}{R} = \frac{49.02}{15} = 3.268\ A$

$$V_{rms} = \frac{V_m}{2}\sqrt{\frac{1}{\pi}\left(\pi - \alpha + \frac{\sin(2\alpha)}{2}\right)}\ \text{at}\ \alpha=66.42°,\ V_{rms}=95.1217V.\ \text{Then},\ I_{rms}=95.122/15=6.34145A$$

$$V_S = \frac{V_m}{\sqrt{2}} = 155.56\ V\ .\ I_S = I_{rms} = 6.34145\ A$$

Then, the rectification efficiency is: $\eta = \frac{P_{dc}}{P_{ac}} = \frac{V_{dc} * I_{dc}}{V_{rms} * I_{rms}} = \frac{49.02 * 3.268}{95.121 * 6.34145} = 26.56\%$

3.3 Single-Phase Full Wave Controlled Rectifier
3.3.1 Single-Phase Center Tap Controlled Rectifier With Resistive Load

Center tap controlled rectifier is shown in Fig.3.8. When the upper half of the transformer secondary is positive and thyristor T1 is triggered, T1 will conduct and the current flows through the load from point a to point b. When the lower half of the transformer secondary is positive and thyristor T2 is triggered, T2 will conduct and the current flows through the load from point a to point b. So, each half of input wave a unidirectional voltage (from a to b) is applied across the load. Various voltages and currents waveforms for center tap controlled rectifier with resistive load are shown in Fig.3.9 and Fig.3.10.

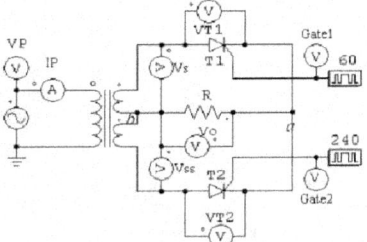

Fig.3.8 Center tap controlled rectifier with resistive load.

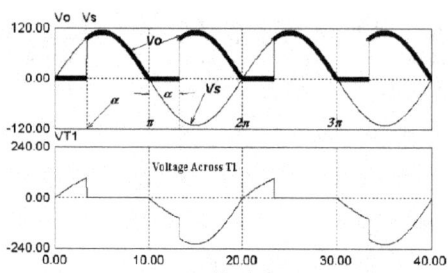

Fig.3.9 The output voltgae and thyristor T1 reverse voltage wavforms along with the supply voltage wavform.

The average voltage, V_{dc}, across the resistive load is given by:

$$V_{dc} = \frac{1}{\pi}\int_{\alpha}^{\pi} V_m \sin(\omega t)\, d\omega t = \frac{V_m}{\pi}(-\cos\pi - \cos(\alpha)) = \frac{V_m}{\pi}(1+\cos\alpha) \quad (3.27)$$

V_{dm} is the maximum output voltage and can be acheaved when $\alpha=0$ in the above equation. The normalized output voltage is: $V_n = \dfrac{V_{dc}}{V_{dm}} = 0.5\,(1+\cos\alpha)$ \quad (3.28)

From the wavfrom of the output voltage shown in Fug.3.9 the rms output voltage can be obtained as following: $V_{rms} = \sqrt{\dfrac{1}{\pi}\int_{\alpha}^{\pi}(V_m \sin(\omega t))^2\, d\omega t} = \dfrac{V_m}{\sqrt{2\pi}}\sqrt{\pi - \alpha + \dfrac{\sin(2\alpha)}{2}}$ \quad (3.29)

Example 4 The rectifier shown in Fig.3.8 has load of $R=15\ \Omega$ and, $V_s=220\ \sin 314\ t$ and unity transformer ratio. If it is required to obtain an average output voltage of 70 % of the maximum possible output voltage, calculate:- (a) The delay angle α. (b) The efficiency. (c) The ripple factor (d) The peak inverse voltage (PIV) of the thyristor.

Solution : (a) V_{dm} is the maximum output voltage and can be acheaved when $\alpha=0$, the normalized output voltage is shown in equation (3.28) which is required to be 70%. Then:

$V_n = \dfrac{V_{dc}}{V_{dm}} = 0.5\,(1+\cos\alpha) = 0.7$, then, $\alpha = 66.42°$

(b) $V_m = 220$, then, $V_{dc} = 0.7*V_{dm} = 0.7*\dfrac{2V_m}{\pi} = 98.04\ V$

$I_{dc} = \dfrac{V_{dc}}{R} = \dfrac{98.04}{15} = 6.536\ A$, $\quad V_{rms} = \dfrac{V_m}{\sqrt{2\pi}}\sqrt{\pi - \alpha + \dfrac{\sin(2\alpha)}{2}}$

at $\alpha=66.42°$ $V_{rms}=134.638\ V$. Then, $I_{rms}=134.638/15=8.976\ A$
$V_S = V_m/\sqrt{2} = 155.56\ V$, $\quad I_S = I_{rms}/\sqrt{2} = 6.347\ A$

Then, The rectification efficiency is: $\eta = \dfrac{P_{dc}}{P_{ac}} = \dfrac{V_{dc}*I_{dc}}{V_{rms}*I_{rms}} = \dfrac{98.04*6.536}{134.638*8.976} = 53.04\%$

3.3.2 Single-Phase Fully Controlled Rectifier Bridge With Resistive Load

This section describes the operation of a single-phase fully-controlled bridge rectifier circuit with resistive load. The operation of this circuit can be understood more easily when the load is pure resistance. The main purpose of the fully controlled bridge rectifier circuit is to provide a variable DC voltage from an AC source.

The circuit of a single-phase fully controlled bridge rectifier circuit is shown in Fig.3.11. The circuit has four SCRs. For this circuit, v_s is a sinusoidal voltage source. When the supply voltage is positive, SCRs T1 and T2 triggered then current flows from v_s through SCR T1, load resistor R (from up to down), SCR T2 and back into the source. In the next half-cycle, the other pair of SCRs T3 and T4 conducts when get pulse on their gates. Then current flows from v_s through SCR T3, load resistor R (from up to down), SCR T4 and back into the source. Even though the direction of current through the source alternates from one

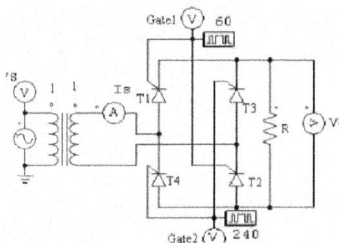

Fig.3.11 Single-phase fully controlled rectifier bridge with resistive load.

half-cycle to the other half-cycle, the current through the load remains unidirectional (from up to down).

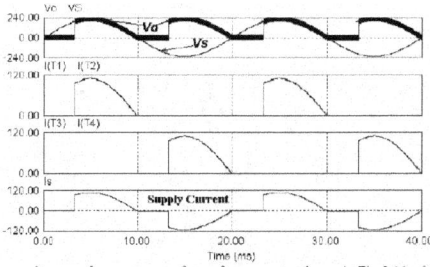

Fig.3.12 Various voltages and currents waveforms for converter shown in Fig.3.11 with resistive load.

HALF WAVE RECTIFIED SOLVED PROBLEMS

Example 20.5. *A half-wave rectifier circuit employing an SCR is adjusted to have a gate current of 1mA. The forward breakdown voltage of SCR is 100 V for I_g = 1mA. If a sinusoidal voltage of 200 V peak is applied, find :*
  (i) *firing angle* (ii) *conduction angle* (iii) *average current.*
Assume load resistance = 100Ω and the holding current to be zero.

Solution. $v = V_m \sin \theta$
  Here. $v = 100$ V. $V_m = 200$ V
(i) ∴ $100 = 200 \sin \theta$
or $\sin \theta = \dfrac{100}{200} = 0.5$
∴ $\theta = \sin^{-1}(0.5) = 30°$ *i.e.* Firing angle, $\alpha = \theta = 30°$
(ii) Conduction angle, $\phi = 180° - \alpha = 180° - 30° = 150°$
(iii) Average voltage $= \dfrac{V_m}{2\pi}(1+\cos \alpha) = \dfrac{200}{2\pi}(1+\cos 30°) = 59.25$ V
∴ Average current $= \dfrac{\text{Average voltage}}{R_L} = \dfrac{59.25}{100} = 0.5925$ A

Example 20.6. *An SCR half-wave rectifier has a forward breakdown voltage of 150 V when a gate current of 1 mA flows in the gate circuit. If a sinusoidal voltage of 400 V peak is applied, find:*
  (i) *firing angle* (ii) *average output voltage*
  (iii) *average current for a load resistance of 200Ω* (iv) *power output*
Assume that the gate current is 1mA throughout and the forward breakdown voltage is more than 400 V when I_g = 1 mA.

Solution. $V_m = 400$ V. $v = 150$ V. $R_L = 200$ Ω
(i) Now $v = V_m \sin \theta$
or $\sin \theta = \dfrac{v}{V_m} = \dfrac{150}{400} = 0.375$
i.e. firing angle, $\alpha (= \theta) = \sin^{-1} 0.375 = 22°$
(ii) Average output voltage is
  $V_{av} = \dfrac{V_m}{2\pi}(1+\cos 22°) = \dfrac{400}{2\pi}(1+\cos 22°) = 122.6$ V
(iii) Average current. $I_{av} = \dfrac{\text{average output voltage}}{R_L} = \dfrac{122.6}{200} = 0.613$ A
(iv) Output power $= V_{av} \times I_{av} = 122.6 \times 0.613 = 75.15$ W

Example 20.7. *An a.c. voltage $v = 240 \sin 314\, t$ is applied to an SCR half-wave rectifier. If the SCR has a forward breakdown voltage of 180 V, find the time during which SCR remains off.*

Solution. The SCR will remain off till the voltage across it reaches 180 V. This is shown in Fig. 20.13. Clearly, SCR will remain off for t second.

Now $\quad v = V_m \sin 314\, t$
Here $\quad v = 180\text{ V}; \quad V_m = 240\text{ V}$
$\therefore \quad 180 = 240 \sin(314\, t)$

or $\quad \sin 314\, t = \dfrac{180}{240} = 0.75$

or $\quad 314\, t = \sin^{-1}(0.75)$
$\qquad\qquad = 48.6° = 0.848 \text{ radian}$

$\therefore \quad t = \dfrac{0.848}{314} = 0.0027 \text{ sec}$
$\qquad\qquad = 2.7 \text{ millisecond}$

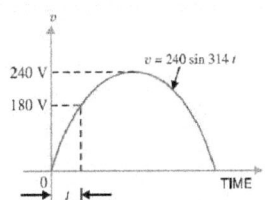

Fig. 20.13

Example 20.8. *In an SCR half-wave rectifier circuit, what peak-load current will occur if we measure an average (d.c.) load current of 1A at a firing angle of 30°?*

Solution. Let I_m be the peak load current.

Now, $\quad I_{av} = \dfrac{V_m}{2\pi R_L}(1 + \cos\alpha)$

$\qquad\qquad = \dfrac{I_m}{2\pi}(1 + \cos\alpha) \qquad (\because I_m = \dfrac{V_m}{R_L})$

$\therefore \quad I_m = \dfrac{2\pi I_{av}}{1 + \cos\alpha}$

Here $\quad I_{av} = I_{dc} = 1\text{A}; \alpha = 30°$

$\therefore \quad I_m = \dfrac{2\pi \times 1}{1 + \cos 30°} = 3.36\text{A}$

Example 20.9. *Power (brightness) of a 100W, 110 V tungsten lamp is to be varied by controlling the firing angle of an SCR in a half-wave rectifier circuit supplied with 110 V a.c. What r.m.s. voltage and current are developed in the lamp at firing angle $\alpha = 60°$?*

Solution. The a.c. voltage is given by:
$$v = V_m \sin \theta$$

Let α be the firing angle as shown in Fig. 20.14. This means that the SCR will fire (*i.e.* start conducting) at $\theta = \alpha$. Clearly, SCR will conduct from α to 180°.

$$E_{r.m.s.}^2 = \frac{1}{2\pi} \int_\alpha^\pi V_m^2 \sin^2 \theta \, d\theta$$

$$= V_m^2 \frac{2(\pi - \alpha) + \sin 2\alpha}{8\pi}$$

$$\therefore \quad E_{r.m.s.} = V_m \sqrt{\frac{2(\pi - \alpha) + \sin 2\alpha}{8\pi}}$$

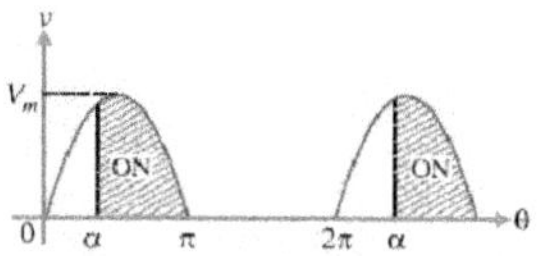

Fig. 20.14

Here, $V_m = \sqrt{2} \times 110 = 156\text{V}; \alpha = 60° = \pi/3$

$$\therefore \quad E_{r.m.s.} = 156 \sqrt{\frac{2(\pi - \pi/3) + \sin 120°}{8\pi}} = 70 \text{ V}$$

Lamp resistance, $R_L = \dfrac{V^2}{P} = \dfrac{(110)^2}{100} = 121 \Omega$

$$\therefore \quad I_{r.m.s.} = \frac{E_{r.m.s.}}{R_L} = \frac{70}{121} = 0.58 \text{ A}$$

FULL WAVE RECTIFIER

Example 20.10. *An SCR full-wave rectifier supplies to a load of 100 Ω. If the peak a.c. voltage between centre tap and one end of secondary is 200V, find (i) d.c. output voltage and (ii) load current for a firing angle of 60°.*

Solution. $V_m = 200$ V; $\alpha = 60°$; $R_L = 100$ Ω

(i) D.C. output voltage, $V_{av} = \dfrac{V_m}{\pi}(1+\cos\alpha) = \dfrac{200}{\pi}(1+\cos 60°) = 95.5$ V

(ii) Load current, $I_{av} = \dfrac{V_{av}}{R_L} = \dfrac{95.5}{100} = 0.955$ A

Example 20.11. *Power (brightness) of a 100 W, 110 V lamp is to be varied by controlling firing angle of SCR full-wave circuit; the r.m.s. value of a.c. voltage appearing across each SCR being 110 V. Find the r.m.s. voltage and current in the lamp at firing angle of 60°.*

Solution. Let $v = v_m \sin\theta$ be the alternating voltage that appears between centre tap and either end of the secondary. Let α be the firing angle as shown in Fig. 20.16. This means that SCR will conduct at $\theta = \alpha$. Clearly, SCR circuit will conduct from α to 180°.

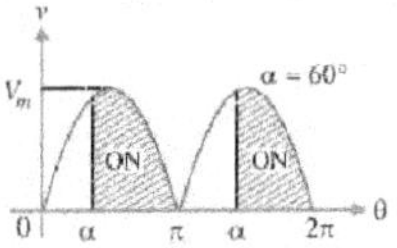

$$E_{r.m.s.}^2 = \dfrac{1}{\pi}\int_{\alpha}^{\pi} V_m^2 \sin^2\theta\, d\theta$$

$$= V_m^2 \dfrac{2(\pi-\alpha)+\sin 2\alpha}{4\pi}$$

$$\therefore \quad E_{r.m.s.} = V_m \sqrt{\dfrac{2(\pi-\alpha)+\sin 2\alpha}{4\pi}}$$

Fig. 20.16

Here $V_m = 110 \times \sqrt{2} = 156$ V; $\alpha = 60°$

$$\therefore \quad E_{r.m.s.} = 156\sqrt{\dfrac{2(\pi-\pi/3)+\sin 120°}{4\pi}} = 98.9\text{V}$$

Lamp resistance, $R_L = \dfrac{V^2}{P} = \dfrac{(110)^2}{100} = 121$ Ω

$$\therefore \quad I_{r.m.s.} = \dfrac{E_{r.m.s.}}{R_L} = \dfrac{98.9}{121} = 0.82 \text{ A}$$

EEL 2003 — Power Electronics — Nasser

Half Wave Uncontrolled Rectifier R Load

Average Voltage		RMS Voltage
$V_{AVG} = V_m / \pi$	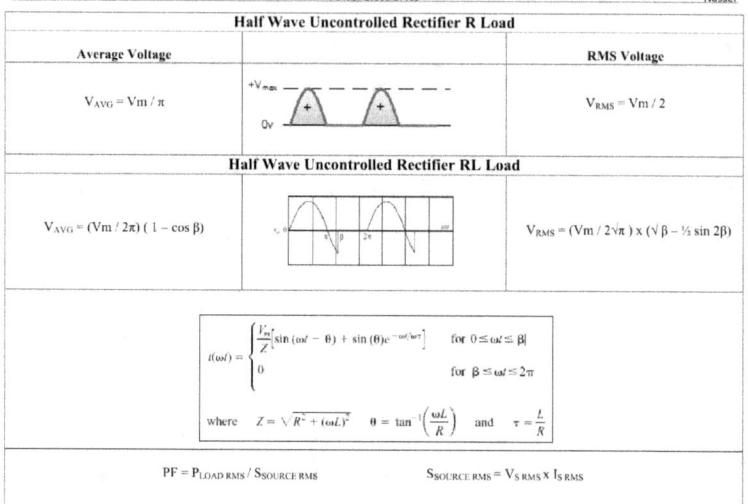	$V_{RMS} = V_m / 2$

Half Wave Uncontrolled Rectifier RL Load

$V_{AVG} = (V_m / 2\pi)(1 - \cos \beta)$		$V_{RMS} = (V_m / 2\sqrt{\pi}) \times (\sqrt{\beta - \tfrac{1}{2} \sin 2\beta})$

$$i(\omega t) = \begin{cases} \dfrac{V_m}{Z}\left[\sin(\omega t - \theta) + \sin(\theta)e^{-\omega t/\omega \tau}\right] & \text{for } 0 \leq \omega t \leq \beta \\ 0 & \text{for } \beta \leq \omega t \leq 2\pi \end{cases}$$

where $\quad Z = \sqrt{R^2 + (\omega L)^2} \quad \theta = \tan^{-1}\left(\dfrac{\omega L}{R}\right) \quad \text{and} \quad \tau = \dfrac{L}{R}$

$PF = P_{LOAD\,RMS} / S_{SOURCE\,RMS}$ $\qquad S_{SOURCE\,RMS} = V_{S\,RMS} \times I_{S\,RMS}$

Half Wave Controlled Rectifier R Load		
Average Voltage		**RMS Voltage**
$V_{AVG} = (V_m / 2\pi)(1 + \cos\alpha)$		$V_{RMS} = (V_m / 2) \times (\sqrt{1 - \alpha/\pi + \sin 2\alpha/2\pi})$
Half Wave Controlled Rectifier RL Load		
$V_{AVG} = (V_m / 2\pi)(\cos\alpha - \cos\beta)$		$V_{RMS} = (V_m / 2) \times (\sqrt{\beta/\pi - \alpha/\pi + \sin 2\beta/2\pi + \sin 2\alpha/2\pi})$
	$i(\omega t) = \begin{cases} \dfrac{V_m}{Z}\left[\sin(\omega t - \theta) - \sin(\alpha - \theta)e^{(\alpha - \omega t)/\omega\tau}\right] & \text{for } \alpha \leq \omega t \leq \beta \\ 0 & \text{otherwise} \end{cases}$	
$PF = P_{LOAD\ RMS} / S_{SOURCE\ RMS}$		$S_{SOURCE\ RMS} = V_{S\ RMS} \times I_{S\ RMS}$

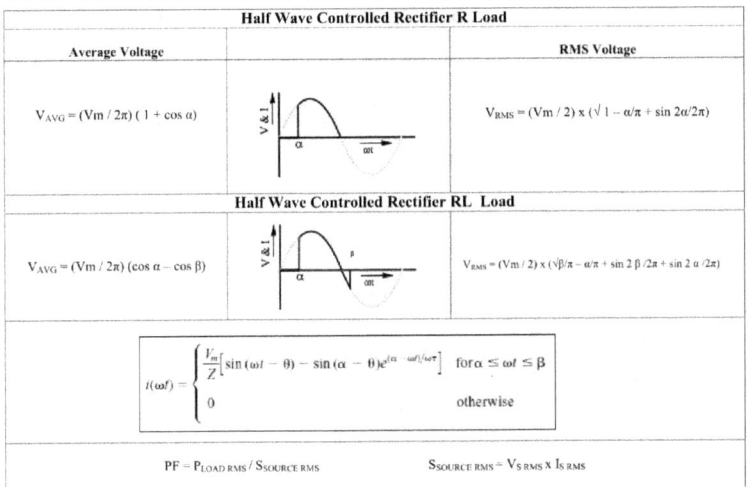

Rectifier Circuit Efficiency

Example 1: The rectifier shown in Fig.2.1 has a pure resistive load of R. Determine (a) The efficiency, (b) Form factor (c) Ripple factor (d) Peak inverse voltage (PIV) of diode D1.
Solution: From Fig 2.2, the average output voltage V_{dc} is defiend as:

$$V_{dc} = \frac{1}{2\pi}\int_0^\pi V_m \sin(\omega t)\, d\omega t = \frac{V_m}{2\pi}(-\cos\pi - \cos(0)) = \frac{V_m}{\pi} \qquad \text{Then, } I_{dc} = \frac{V_{dc}}{R} = \frac{V_m}{\pi R}$$

$$V_{rms} = \sqrt{\frac{1}{2\pi}\int_0^\pi (V_m \sin \omega t)^2} = \frac{V_m}{2}, \quad I_{rms} = \frac{V_m}{2R} \text{ and } V_S = \frac{V_m}{\sqrt{2}}$$

The rms value of the transformer secondery current is the same as that of the load: $I_S = V_m/2R$
Then, the efficiency or rectification ratio is:

$$\eta = \frac{P_{dc}}{P_{ac}} = \frac{V_{dc}*I_{dc}}{V_{rms}*I_{rms}} = \frac{\frac{V_m}{\pi}*\frac{V_m}{\pi R}}{\frac{V_m}{2}*\frac{V_m}{2R}} = 40.53\%$$

Example 4 The rectifier shown in Fig.2.12 has a purely resistive load of $R=15\,\Omega$ and, $V_S=300\sin 314\,t$ and unity transformer ratio. Determine (a) The efficiency, (b) Form factor, (c) Ripple factor, (d) The peak inverse voltage, (PIV) of each diode, , and, (e) Input power factor.
Solution: $V_m = 300$ V

$$V_{dc} = \frac{1}{\pi}\int_0^\pi V_m \sin \omega t\, d\omega t = \frac{2V_m}{\pi} = 190.956\ V, \quad I_{dc} = \frac{2V_m}{\pi R} = 12.7324\ A$$

$$V_{rms} = \left[\frac{1}{\pi}\int_0^\pi (V_m \sin \omega t)^2\, d\omega t\right]^{1/2} = \frac{V_m}{\sqrt{2}} = 212.132\ V, \quad I_{rms} = \frac{V_m}{\sqrt{2}\,R} = 14.1421$$

(a) $\eta = \dfrac{P_{dc}}{P_{ac}} = \dfrac{V_{dc}\,I_{dc}}{V_{rms}\,I_{rms}} = 81.06\ \%$

(b) $FF = \dfrac{V_{rms}}{V_{dc}} = 1.11$

(c) $RF = \dfrac{V_{ac}}{V_{dc}} = \dfrac{\sqrt{V_{rms}^2 - V_{dc}^2}}{V_{dc}} = \sqrt{\dfrac{V_{rms}^2}{V_{dc}^2} - 1} = \sqrt{FF^2 - 1} = 0.482$

(d) The PIV=$300V$

(e) Input power factor $= \dfrac{Real\ Power}{Apperant\ Power} = \dfrac{V_S\,I_S\cos\phi}{V_S\,I_S} = 1$

D. Part 4: Investigate the DC-to-DC power conversion circuits used in power applications.

1. State the principle of step-down and step-up operations.
2. Explain the DC chopper classification and describe switch-mode regulators
3. Explain the operation of buck, boost
4. Explain the operation buck-boost regulators.

CHAPTER 6

DC-DC Converters

Dc-dc converters are power electronic circuits that convert a dc voltage to a different dc voltage level, often providing a regulated output. The circuits described in this chapter are classified as switched-mode dc-dc converters, also called switching power supplies or switchers. This chapter describes some basic dc-dc converter circuits. Chapter 7 describes some common variations of these circuits that are used in many dc power supply designs.

6.1 LINEAR VOLTAGE REGULATORS

Before we discuss switched-mode converters, it is useful to review the motivation for an alternative to linear dc-dc converters that was introduced in Chapt. 1. One method of converting a dc voltage to a lower dc voltage is a simple circuit as shown in Fig. 6-1. The output voltage is

$$V_o = I_L R_L$$

where the load current is controlled by the transistor. By adjusting the transistor base current, the output voltage may be controlled over a range of 0 to roughly V_s. The base current can be adjusted to compensate for variations in the supply voltage or the load, thus regulating the output. This type of circuit is called a linear dc-dc converter or a linear regulator because the transistor operates in the linear region, rather than in the saturation or cutoff regions. The transistor in effect operates as a variable resistance.

While this may be a simple way of converting a dc supply voltage to a lower dc voltage and regulating the output, the low efficiency of this circuit is a serious drawback for power applications. The power absorbed by the load is $V_o I_L$, and

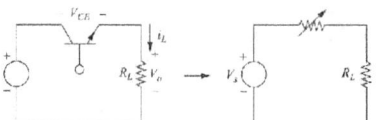

Figure 6-1 A basic linear regulator.

the power absorbed by the transistor is $V_{CE}I_L$, assuming a small base current. The power loss in the transistor makes this circuit inefficient. For example, if the output voltage is one-quarter of the input voltage, the load resistor absorbs one-quarter of the source power, which is an efficiency of 25 percent. The transistor absorbs the other 75 percent of the power supplied by the source. Lower output voltages result in even lower efficiencies. Therefore, the linear voltage regulator is suitable only for low-power applications.

6.2 A BASIC SWITCHING CONVERTER

An efficient alternative to the linear regulator is the switching converter. In a switching converter circuit, the transistor operates as an electronic switch by being completely on or completely off (saturation or cutoff for a BJT or the triode and cutoff regions of a MOSFET). This circuit is also known as a dc chopper.

Assuming the switch is ideal in Fig. 6-2, the output is the same as the input when the switch is closed, and the output is zero when the switch is open. Periodic

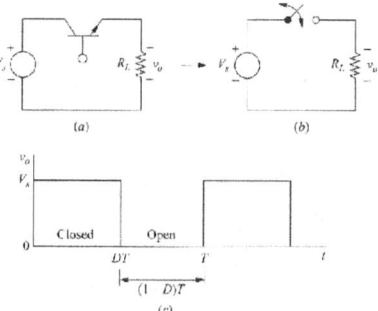

Figure 6-2 (a) A basic dc-dc switching converter; (b) Switching equivalent; (c) Output voltage.

opening and closing of the switch results in the pulse output shown in Fig. 6-2c. The average or dc component of the output voltage is

$$V_o = \frac{1}{T}\int_0^T v_o(t)dt = \frac{1}{T}\int_0^{DT} V_s dt = V_s D \qquad (6\text{-}1)$$

The dc component of the output voltage is controlled by adjusting the duty ratio D, which is the fraction of the switching period that the switch is closed

$$D \equiv \frac{t_{on}}{t_{on} + t_{off}} = \frac{t_{on}}{T} = t_{on}f \qquad (6\text{-}2)$$

where f is the switching frequency. The dc component of the output voltage will be less than or equal to the input voltage for this circuit.

The power absorbed by the ideal switch is zero. When the switch is open, there is no current in it; when the switch is closed, there is no voltage across it. Therefore, all power is absorbed by the load, and the energy efficiency is 100 percent. Losses will occur in a real switch because the voltage across it will not be zero when it is on, and the switch must pass through the linear region when making a transition from one state to the other.

6.3 THE BUCK (STEP-DOWN) CONVERTER

Controlling the dc component of a pulsed output voltage of the type in Fig. 6-2c may be sufficient for some applications, such as controlling the speed of a dc motor, but often the objective is to produce an output that is purely dc. One way of obtaining a dc output from the circuit of Fig. 6-2a is to insert a low-pass filter after the switch. Figure 6-3a shows an LC low-pass filter added to the basic converter. The diode provides a path for the inductor current when the switch is opened and is reverse-biased when the switch is closed. This circuit is called a *buck converter* or a *step-down converter* because the output voltage is less than the input.

Voltage and Current Relationships

If the low-pass filter is ideal, the output voltage is the average of the input voltage to the filter. The input to the filter, v_x in Fig. 6-3a, is V_s when the switch is closed and is zero when the switch is open, provided that the inductor current remains positive, keeping the diode on. If the switch is closed periodically at a duty ratio D, the average voltage at the filter input is $V_s D$, as in Eq. (6-1).

This analysis assumes that the diode remains forward-biased for the entire time when the switch is open, implying that the inductor current remains positive. An inductor current that remains positive throughout the switching period is known as *continuous current*. Conversely, discontinuous current is characterized by the inductor current's returning to zero during each period.

6.3 The Buck (Step-Down) Converter

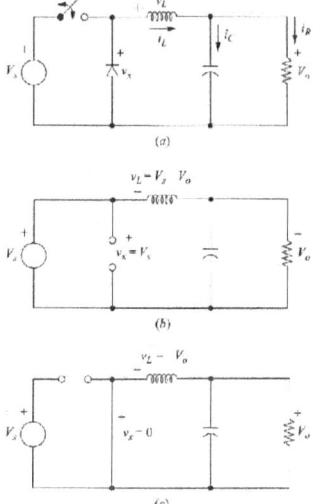

Figure 6-3 (a) Buck dc-dc converter; (b) Equivalent circuit for the switch closed; (c) Equivalent circuit for the switch open.

Another way of analyzing the operation of the buck converter of Fig. 6-3a is to examine the inductor voltage and current. This analysis method will prove useful for designing the filter and for analyzing circuits that are presented later in this chapter.

Buck converters and dc-dc converters in general, have the following properties when operating in the steady state:

1. The inductor current is periodic.

$$i_L(t+T) = i_L(t) \qquad (6\text{-}3)$$

2. The average inductor voltage is zero (see Sec. 2.3).

$$V_L = \frac{1}{T} \int_t^{t+T} v_L(\lambda)d\lambda = 0 \qquad (6\text{-}4)$$

3. The average capacitor current is zero (see Sec. 2.3).

$$I_C = \frac{1}{T} \int_t^{t+T} i_C(\lambda)d\lambda = 0 \qquad (6\text{-}5)$$

4. The power supplied by the source is the same as the power delivered to the load. For nonideal components, the source also supplies the losses.

$$\begin{aligned} P_s &= P_o & \text{ideal} \\ P_s &= P_o + \text{losses} & \text{nonideal} \end{aligned} \qquad (6\text{-}6)$$

Analysis of the buck converter of Fig. 6-3a begins by making these assumptions:

1. The circuit is operating in the steady state.
2. The inductor current is continuous (always positive).
3. The capacitor is very large, and the output voltage is held constant at voltage V_o. This restriction will be relaxed later to show the effects of finite capacitance.
4. The switching period is T; the switch is closed for time DT and open for time $(1-D)T$.
5. The components are ideal.

The key to the analysis for determining the output V_o is to examine the inductor current and inductor voltage first for the switch closed and then for the switch open. The net change in inductor current over one period must be zero for steady-state operation. The average inductor voltage is zero.

Analysis for the Switch Closed When the switch is closed in the buck converter circuit of Fig. 6-3a, the diode is reverse-biased and Fig. 6-3b is an equivalent circuit. The voltage across the inductor is

$$v_L = V_s - V_o = L\frac{di_L}{dt}$$

Rearranging,

$$\frac{di_L}{dt} = \frac{V_s - V_o}{L} \qquad \text{switch closed}$$

Since the derivative of the current is a positive constant, the current increases linearly as shown in Fig. 6-4b. The change in current while the switch is closed is computed by modifying the preceding equation.

$$\frac{di_L}{dt} = \frac{\Delta i_L}{\Delta t} = \frac{\Delta i_L}{DT} = \frac{V_s - V_o}{L}$$

$$(\Delta i_L)_{\text{closed}} = \left(\frac{V_s - V_o}{L}\right)DT \qquad (6\text{-}7)$$

6.3 The Buck (Step-Down) Converter

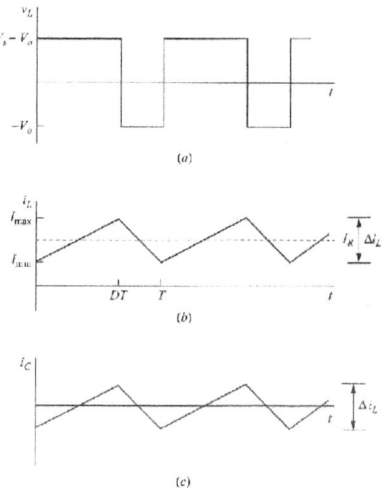

Figure 6-4 Buck converter waveforms: (*a*) Inductor voltage; (*b*) Inductor current; (*c*) Capacitor current.

Analysis for the Switch Open When the switch is open, the diode becomes forward-biased to carry the inductor current and the equivalent circuit of Fig. 6-3c applies. The voltage across the inductor when the switch is open is

$$v_L = -V_o = L\frac{di_L}{dt}$$

Rearranging,

$$\frac{di_L}{dt} = \frac{-V_o}{L} \quad \text{switch open}$$

The derivative of current in the inductor is a negative constant, and the current decreases linearly as shown in Fig. 6-4b. The change in inductor current when the switch is open is

$$\frac{\Delta i_L}{\Delta t} = \frac{\Delta i_L}{(1-D)T} = -\frac{V_o}{L}$$

$$(\Delta i_L)_{\text{open}} = -\left(\frac{V_o}{L}\right)(1-D)T \qquad (6\text{-}8)$$

Steady-state operation requires that the inductor current at the end of the switching cycle be the same as that at the beginning, meaning that the net change in inductor current over one period is zero. This requires

$$(\Delta i_L)_{closed} + (\Delta i_L)_{open} = 0$$

Using Eqs. (6-7) and (6-8),

$$\left(\frac{V_s - V_o}{L}\right)(DT) - \left(\frac{V_o}{L}\right)(1 - D)T = 0$$

Solving for V_o,

$$\boxed{V_o = V_s D} \qquad (6\text{-}9)$$

which is the same result as Eq. (6-1). *The buck converter produces an output voltage that is less than or equal to the input.*

An alternative derivation of the output voltage is based on the inductor voltage, as shown in Fig. 6-4a. Since the average inductor voltage is zero for periodic operation,

$$V_L = (V_s - V_o)DT + (-V_o)(1 - D)T = 0$$

Solving the preceding equation for V_o yields the same result as Eq. (6-9), $V_o = V_s D$.

Note that the output voltage depends on only the input and the duty ratio D. If the input voltage fluctuates, the output voltage can be regulated by adjusting the duty ratio appropriately. A feedback loop is required to sample the output voltage, compare it to a reference, and set the duty ratio of the switch accordingly. Regulation techniques are discussed in Chap. 7.

The average inductor current must be the same as the average current in the load resistor, since the average capacitor current must be zero for steady-state operation:

$$I_L = I_R = \frac{V_o}{R} \qquad (6\text{-}10)$$

Since the change in inductor current is known from Eqs. (6-7) and (6-8), the maximum and minimum values of the inductor current are computed as

$$\begin{aligned} I_{max} &= I_L + \frac{\Delta i_L}{2} \\ &= \frac{V_o}{R} + \frac{1}{2}\left[\frac{V_o}{L}(1-D)T\right] = V_o\left(\frac{1}{R} + \frac{1-D}{2Lf}\right) \end{aligned} \qquad (6\text{-}11)$$

$$\begin{aligned} I_{min} &= I_L - \frac{\Delta i_L}{2} \\ &= \frac{V_o}{R} - \frac{1}{2}\left[\frac{V_o}{L}(1-D)T\right] = V_o\left(\frac{1}{R} - \frac{1-D}{2Lf}\right) \end{aligned} \qquad (6\text{-}12)$$

where $f = 1/T$ is the switching frequency.

For the preceding analysis to be valid, continuous current in the inductor must be verified. An easy check for continuous current is to calculate the minimum inductor current from Eq. (6-12). Since the minimum value of inductor current must be positive for continuous current, a negative minimum calculated from Eq. (6-12) is not allowed due to the diode and indicates discontinuous current. The circuit will operate for discontinuous inductor current, but the preceding analysis is not valid. Discontinuous-current operation is discussed later in this chapter.

Equation (6-12) can be used to determine the combination of L and f that will result in continuous current. Since $I_{\min} = 0$ is the boundary between continuous and discontinuous current,

$$I_{\min} = 0 = V_o\left(\frac{1}{R} - \frac{1-D}{2Lf}\right)$$

$$(Lf)_{\min} = \frac{(1-D)R}{2} \qquad (6\text{-}13)$$

If the desired switching frequency is established,

$$L_{\min} = \frac{(1-D)R}{2f} \quad \text{for continuous current} \qquad (6\text{-}14)$$

where $L_{\min}$ is the minimum inductance required for continuous current. In practice, a value of inductance greater than $L_{\min}$ is desirable to ensure continuous current.

In the design of a buck converter, the peak-to-peak variation in the inductor current is often used as a design criterion. Equation (6-7) can be combined with Eq. (6-9) to determine the value of inductance for a specified peak-to-peak inductor current for continuous-current operation:

$$\Delta i_L = \left(\frac{V_s - V_o}{L}\right)DT = \left(\frac{V_s - V_o}{Lf}\right)D = \frac{V_o(1-D)}{Lf} \qquad (6\text{-}15)$$

or

$$L = \left(\frac{V_s - V_o}{\Delta i_L f}\right)D = \frac{V_o(1-D)}{\Delta i_L f} \qquad (6\text{-}16)$$

Since the converter components are assumed to be ideal, the power supplied by the source must be the same as the power absorbed by the load resistor.

$$P_s = P_o$$
$$V_s I_s = V_o I_o \qquad (6\text{-}17)$$

or

$$\frac{V_o}{V_s} = \frac{I_s}{I_o}$$

Note that the preceding relationship is similar to the voltage-current relationship for a transformer in ac applications. Therefore, the buck converter circuit is equivalent to a dc transformer.

Output Voltage Ripple

In the preceding analysis, the capacitor was assumed to be very large to keep the output voltage constant. In practice, the output voltage cannot be kept perfectly constant with a finite capacitance. The variation in output voltage, or ripple, is computed from the voltage-current relationship of the capacitor. The current in the capacitor is

$$i_C = i_L - i_R$$

shown in Fig. 6-5a.

While the capacitor current is positive, the capacitor is charging. From the definition of capacitance,

$$Q = CV_o$$
$$\Delta Q = C \Delta V_o$$
$$\Delta V_o = \frac{\Delta Q}{C}$$

The change in charge ΔQ is the area of the triangle above the time axis

$$\Delta Q = \frac{1}{2} \left(\frac{T}{2} \right) \left(\frac{\Delta i_L}{2} \right) = \frac{T \Delta i_L}{8}$$

resulting in

$$\Delta V_o = \frac{T \Delta i_L}{8C}$$

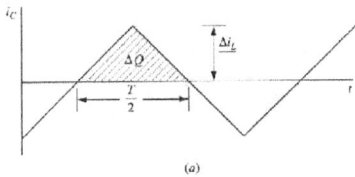

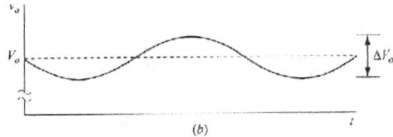

Figure 6-5 Buck converter waveforms. (a) Capacitor current; (b) Capacitor ripple voltage.

Using Eq. (6-8) for Δi_L,

$$\Delta V_o = \frac{TV_o}{8CL}(1-D)T = \frac{V_o(1-D)}{8LCf^2} \qquad (6\text{-}18)$$

In this equation, ΔV_o is the peak-to-peak ripple voltage at the output, as shown in Fig. 6-5b. It is also useful to express the ripple as a fraction of the output voltage,

$$\boxed{\frac{\Delta V_o}{V_o} = \frac{1-D}{8LCf^2}} \qquad (6\text{-}19)$$

In design, it is useful to rearrange the preceding equation to express required capacitance in terms of specified voltage ripple:

$$\boxed{C = \frac{1-D}{8L(\Delta V_o/V_o)f^2}} \qquad (6\text{-}20)$$

If the ripple is not large, the assumption of a constant output voltage is reasonable and the preceding analysis is essentially valid.

EXAMPLE 6-1

Buck Converter

The buck dc-dc converter of Fig. 6-3a has the following parameters:

$V_s = 50$ V
$D = 0.4$
$L = 400$ μH
$C = 100$ μF
$f = 20$ kHz
$R = 20$ Ω

Assuming ideal components, calculate (a) the output voltage V_o, (b) the maximum and minimum inductor current, and (c) the output voltage ripple.

■ **Solution**

(a) The inductor current is assumed to be continuous, and the output voltage is computed from Eq. (6-9),

$$V_o = V_s D = (50)(0.4) = 20 \text{ V}$$

(b) Maximum and minimum inductor currents are computed from Eqs. (6-11) and (6-12).

$$I_{max} = V_o\left(\frac{1}{R} + \frac{1-D}{2Lf}\right)$$

$$= 20\left[\frac{1}{20} + \frac{1-0.4}{2(400)(10)^{-6}(20)(10)^3}\right]$$

$$= 1 + \frac{1.5}{2} = 1.75 \text{ A}$$

CHAPTER 6 DC-DC Converters

$$I_{max} = V_o\left(\frac{1}{R} - \frac{1-D}{2Lf}\right)$$

$$= 1 - \frac{1.5}{2} = 0.25 \text{ A}$$

The average inductor current is 1 A, and $\Delta i_L = 1.5$ A. Note that the minimum inductor current is positive, verifying that the assumption of continuous current was valid.

(c) The output voltage ripple is computed from Eq. (6-19).

$$\frac{\Delta V_o}{V_o} = \frac{1-D}{8LCf^2} = \frac{1-0.4}{8(400)(10)^{-6}(100)(10)^{-6}(20,000)^2}$$

$$= 0.00469 = 0.469\%$$

Since the output ripple is sufficiently small, the assumption of a constant output voltage was reasonable.

Capacitor Resistance—The Effect on Ripple Voltage

The output voltage ripple in Eqs. (6-18) and (6-19) is based on an ideal capacitor. A real capacitor can be modeled as a capacitance with an equivalent series resistance (ESR) and an equivalent series inductance (ESL). The ESR may have a significant effect on the output voltage ripple, often producing a ripple voltage greater than that of the ideal capacitance. The inductance in the capacitor is usually not a significant factor at typical switching frequencies. Figure 6.6 shows a capacitor model that is appropriate for most applications.

The ripple due to the ESR can be approximated by first determining the current in the capacitor, assuming the capacitor to be ideal. For the buck converter in the continuous-current mode, capacitor current is the triangular current waveform of Fig. 6-4c. The voltage variation across the capacitor resistance is

$$\Delta V_{o,\text{ESR}} = \Delta i_C r_C = \Delta i_L r_C \qquad (6\text{-}21)$$

To estimate a worst-case condition, one could assume that the peak-to-peak ripple voltage due to the ESR algebraically adds to the ripple due to the capacitance. However, the peaks of the capacitor and the ESR ripple voltages will not coincide, so

$$\Delta V_o < \Delta V_{o,C} + \Delta V_{o,\text{ESR}} \qquad (6\text{-}22)$$

where $\Delta V_{o,C}$ is ΔV_o in Eq. (6-18). The ripple voltage due to the ESR can be much larger than the ripple due to the pure capacitance. In that case, the output capacitor is chosen on the basis of the equivalent series resistance rather than capacitance only.

$$\Delta V_o \approx \Delta V_{o,\text{ESR}} = \Delta i_C r_C \qquad (6\text{-}23)$$

Figure 6-6 A model for the capacitor including the equivalent series resistance (ESR).

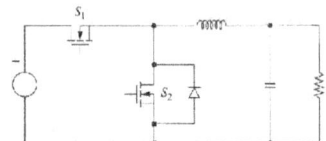

Figure 6-7 A synchronous buck converter. The MOSFET S_2 carries the inductor current when S_1 is off to provide a lower voltage drop than a diode.

between small component size and efficiency. Other designers prefer to use lower switching frequencies of about 50 kHz to keep switching losses small, while still others prefer frequencies larger than 1 MHz. As switching devices improve, switching frequencies will increase.

For low-voltage, high-current applications, the synchronous rectification scheme of Fig. 6-7 is preferred over using a diode for the second switch. The voltage across the conducting MOSFET will be much less than that across a diode, resulting in lower losses.

The inductor value should be larger than L_{min} in Eq. (6-14) to ensure continuous-current operation. Some designers select a value 25 percent larger than L_{min}. Other designers use different criteria, such as setting the inductor current variation, Δi_L in Eq. (6-15), to a desired value, such as 40 percent of the average inductor current. A smaller Δi_L results in lower peak and rms inductor currents and a lower rms capacitor current but requires a larger inductor.

The inductor wire must be rated at the rms current, and the core should not saturate for peak inductor current. The capacitor must be selected to limit the output ripple to the design specifications, to withstand peak output voltage, and to carry the required rms current.

The switch (usually a MOSFET with a low $R_{DS_{on}}$) and diode (or second MOSFET for synchronous rectification) must withstand maximum voltage stress when off and maximum current when on. The temperature ratings must not be exceeded, often requiring a heat sink.

Assuming ideal switches and an ideal inductor in the initial design is usually reasonable. However, the ESR of the capacitor should be included because it typically gives a more significant output voltage ripple than the ideal device and greatly influences the choice of capacitor size.

EXAMPLE 6-2

Buck Converter Design 1

Design a buck converter to produce an output voltage of 18 V across a 10-Ω load resistor. The output voltage ripple must not exceed 0.5 percent. The dc supply is 48 V. Design for continuous inductor current. Specify the duty ratio, the switching frequency, the values of the inductor and capacitor, the peak voltage rating of each device, and the rms current in the inductor and capacitor. Assume ideal components.

■ Solution

Using the buck converter circuit in Fig. 6-3a, the duty ratio for continuous-current operation is determined from Eq. (6-9):

$$D = \frac{V_o}{V_s} = \frac{18}{48} = 0.375$$

The switching frequency and inductor size must be selected for continuous-current operation. Let the switching frequency arbitrarily be 40 kHz, which is well above the audio range and is low enough to keep switching losses small. The minimum inductor size is determined from Eq. (6-14):

$$L_{min} = \frac{(1-D)(R)}{2f} = \frac{(1-0.375)(10)}{2(40,000)} = 78\,\mu H$$

Let the inductor be 25 percent larger than the minimum to ensure that inductor current is continuous.

$$L = 1.25 L_{min} = (1.25)(78\,\mu H) = 97.5\,\mu H$$

Average inductor current and the change in current are determined from Eqs. (6-10) and (6-17).

$$I_L = \frac{V_o}{R} = \frac{18}{10} = 1.8\,A$$

$$\Delta i_L = \left(\frac{V_s - V_o}{L}\right) DT = \frac{48 - 18}{97.5(10)^{-6}}(0.375)\left(\frac{1}{40,000}\right) = 2.88\,A$$

The maximum and minimum inductor currents are determined from Eqs. (6-11) and (6-12).

$$I_{max} = I_L + \frac{\Delta i_L}{2} = 1.8 + 1.44 = 3.24\,A$$

$$I_{min} = I_L - \frac{\Delta i_L}{2} = 1.8 - 1.44 = 0.36\,A$$

The inductor must be rated for rms current, which is computed as in Chap. 2 (see Example 2-8). For the offset triangular wave,

$$I_{L,rms} = \sqrt{I_L^2 + \left(\frac{\Delta i_L/2}{\sqrt{3}}\right)^2} = \sqrt{(1.8)^2 + \left(\frac{1.44}{\sqrt{3}}\right)^2} = 1.98\,A$$

The capacitor is selected using Eq. (6-20).

$$C = \frac{1-D}{8L(\Delta V_o/V_o)f^2} = \frac{1-0.375}{8(97.5)(10)^{-6}(0.005)(40,000)^2} = 100\,\mu F$$

Peak capacitor current is $\Delta i_L/2 = 1.44\,A$, and rms capacitor current for the triangular waveform is $1.44/\sqrt{3} = 0.83\,A$. The maximum voltage across the switch and diode is V_s, or 48 V. The inductor voltage when the switch is closed is $V_s - V_o = 48 - 18 = 30$ V. The inductor voltage when the switch is open is $V_o = 18$ V. Therefore, the inductor must withstand 30 V. The capacitor must be rated for the 18-V output.

EXAMPLE 6-3

Buck Converter Design 2

Power supplies for telecommunications applications may require high currents at low voltages. Design a buck converter that has an input voltage of 3.3 V and an output voltage of 1.2 V. The output current varies between 4 and 6 A. The output voltage ripple must not exceed 2 percent. Specify the inductor value such that the peak-to-peak variation in inductor current does not exceed 40 percent of the average value. Determine the required rms current rating of the inductor and of the capacitor. Determine the maximum equivalent series resistance of the capacitor.

■ **Solution**

Because of the low voltage and high output current in this application, the synchronous rectification buck converter of Fig. 6-7 is used. The duty ratio is determined from Eq. (6-9).

$$D = \frac{V_o}{V_s} = \frac{1.2}{3.3} = 0.364$$

The switching frequency and inductor size must be selected for continuous-current operation. Let the switching frequency arbitrarily be 500 kHz to give a good tradeoff between small component size and low switching losses.

The average inductor current is the same as the output current. Analyzing the circuit for an output current of 4 A,

$$I_L = I_o = 4 \text{ A}$$

$$\Delta i_L = (40\%)(4) = 1.6 \text{ A}$$

Using Eq. (6-16),

$$L = \left(\frac{V_s - V_o}{\Delta i_L f}\right)D = \frac{3.3 - 1.2}{(1.6)(500{,}000)}(0.364) = 0.955 \text{ } \mu\text{H}$$

Analyzing the circuit for an output current of 6 A,

$$I_L = I_o = 6 \text{ A}$$

$$\Delta i_L = (40\%)(6) = 2.4 \text{ A}$$

resulting in

$$L = \left(\frac{V_s - V_o}{\Delta i_L f}\right)D = \frac{3.3 - 1.2}{(2.4)(500{,}000)}(0.364) = 0.636 \text{ } \mu\text{H}$$

Since 0.636 μH would be too small for the 4-A output, use $L = 0.955$ μH, which would be rounded to 1 μH.

Inductor rms current is determined from

$$I_{L,\text{rms}} = \sqrt{I_L^2 + \left(\frac{\Delta i_L/2}{\sqrt{3}}\right)^2}$$

(see Chap. 2). From Eq. (6-15), the variation in inductor current is 1.6 A for each output current. Using the 6-A output current, the inductor must be rated for an rms current of

$$I_{L,\text{rms}} = \sqrt{6^2 + \left(\frac{0.8}{\sqrt{3}}\right)^2} = 6.02 \text{ A}$$

Note that the average inductor current would be a good approximation to the rms current since the variation is relatively small.

Using $L = 1$ μH in Eq. (6-20), the minimum capacitance is determined as

$$C = \frac{1-D}{8L(\Delta V_o/V_o)f^2} = \frac{1-0.364}{8(1)(10)^{-6}(0.02)(500,000)^2} = 0.16 \text{ μF}$$

The allowable output voltage ripple of 2 percent is $(0.02)(1.2) = 24$ mV. The maximum ESR is computed from Eq. (6-23).

$$\Delta V_o \approx r_C \Delta i_C = r_C \Delta i_L$$

or

$$r_C = \frac{\Delta V_o}{\Delta i_C} = \frac{24 \text{ mV}}{1.6 \text{ A}} = 15 \text{ m}\Omega$$

At this point, the designer would search manufacturer's specifications for a capacitor having 15-mΩ ESR. The capacitor may have to be much larger than the calculated value of 0.16 μF to meet the ESR requirement. Peak capacitor current is $\Delta i_L/2 = 0.8$ A, and rms capacitor current for the triangular waveform is $0.8/\sqrt{3} = 0.46$ A.

6.5 THE BOOST CONVERTER

The boost converter is shown in Fig. 6-8. This is another switching converter that operates by periodically opening and closing an electronic switch. It is called a boost converter because the output voltage is larger than the input.

Voltage and Current Relationships

The analysis assumes the following:

1. Steady-state conditions exist.
2. The switching period is T, and the switch is closed for time DT and open for $(1-D)T$.
3. The inductor current is continuous (always positive).
4. The capacitor is very large, and the output voltage is held constant at voltage V_o.
5. The components are ideal.

The analysis proceeds by examining the inductor voltage and current for the switch closed and again for the switch open.

DC-DC Converters

➢ The purpose of a DC-DC converter is to supply a regulated DC output voltage to a variable-load resistance from a fluctuating DC input voltage.

➢ In many cases the DC input voltage is obtained by rectifying a line voltage that is changing in magnitude.

➢ DC-DC converters are commonly used in applications requiring regulated DC power, such as :
        1- Computers,
        2- Medical instrumentation,
        3- Communication devices,
        4- Television receivers,
        5- Battery chargers, and
        6- DC motor speed control applications.

> The output voltage in DC-DC converters is generally controlled using a switching concept, as illustrated by the basic DC-DC converter shown in Fig. 1.

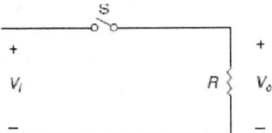

Fig. 2. Basic DC-DC converter.

> Early DC-DC converters were known as choppers with silicon-controlled rectifiers (SCRs) used as the switching mechanisms.

> Modern DC-DC converters classified as switch mode power supplies (SMPS) employ insulated gate bipolar transistors (IGBTs) and metal oxide silicon field effect transistors (MOSFETs).

- The switch mode power supply has several functions :

    1. Step down an unregulated DC input voltage to produce a regulated DC output voltage using a buck or step-down converter.

    2. Step up an unregulated DC input voltage to produce a regulated DC output voltage using a boost or step-up converter.

    3. Step down and then step up an unregulated DC input voltage to produce a regulated DC output voltage using a buck–boost converter.

➢ The regulation of the average output voltage in a DC-DC converter is a function of the on-time t_{on} of the switch, the pulse width, and the switching frequency f_s as illustrated in Fig. 2.

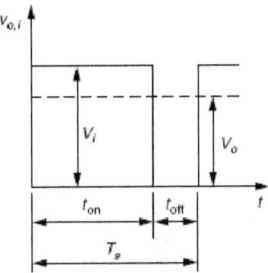

Fig. 2. DC-DC converter voltage waveforms.

▪ Pulse width modulation (PWM) is the most widely used method of controlling the output voltage. The PWM concept is illustrated in Fig. 3.

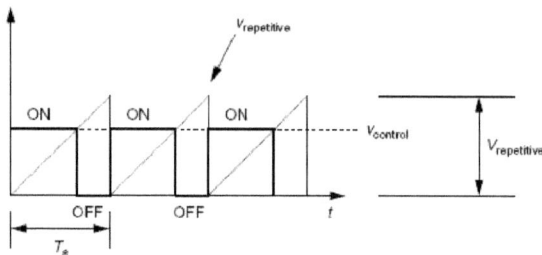

Fig.3. Pulse width modulation concept.

➢ The output voltage control depends on the duty ratio D. The duty ratio is defined as:
$$D = \frac{t_{on}}{T_s} = \frac{V_{control}}{V_{repetitive}}$$

➤ Based on the on-time t_{on} of the switch and the switching period T_s. PWM switching involves comparing the level of a control voltage $v_{control}$ to the level of a repetitive waveform as illustrated in Fig. 3.

▪ The on-time of the switch is defined as the portion of the switching period where the value of the repetitive waveform is less than the control voltage.

➤ The switching period (switching frequency) remains constant while the control voltage level is adjusted to change the on-time and therefore the duty ratio of the switch. The switching frequency is usually chosen above 20 kHz so the noise is outside the audio range.

➤ DC-DC converters operate in one of two modes depending on the characteristics of the output current:

        1. Continuous conduction
        2. Discontinuous conduction

- The continuous-conduction mode is defined by continuous output current (greater than zero) over the entire switching period, whereas the discontinuous conduction mode is defined by discontinuous output current (equal to zero) during any portion of the switching period.

Switched Mode Power Supplies (SMPS)

- DC chopper can be used as switching-mode power supplies to convert unregulated input dc voltage to regulated output dc voltage.

- The regulation is normally achieved by PWM at fixed frequency.

- The switching device is usually a power BJT, MOSFET or IGBT.

- The ripple content of the output voltage is reduced through LC filters.

- There are 4-basic topologies of switching regulators:

    1- Buck regulators,

    2- Boost regulators,

    3- Buck-Boost regulators, and

    4- Cuke regulators.

1- Buck Regulator

➢ The buck regulator is a step down chopper, has an average voltage less than the input voltage.

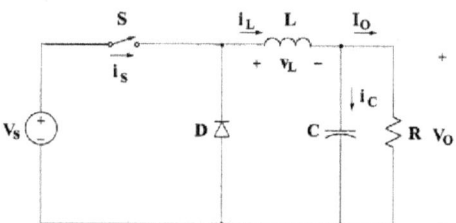

▢ There are 2-modes of operation, mode (1) when the transistor (switch S) is ON or closed, and mode (2) when the transistor (switch S) is OFF or open.

Mode (1)

⏋ When S is on at $t=0$

▸ Input current flows through filter inductor (L), filter capacitor (C) and load R.

$$e_L = L \frac{di}{dt}$$

➤ Assuming i_L rises from I_1 to I_2 in time t_1.

$$e_L = L \frac{(I_2 - I_1)}{(t_1 - 0)} = L \frac{I_2 - I_1}{t_1}$$

$$\because e_L = V_s - V_o = L \frac{I_2 - I_1}{t_1} = L \frac{\Delta I}{t_1}$$

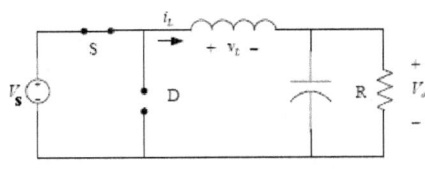

CIRCUIT WHEN SWITCH IS CLOSED

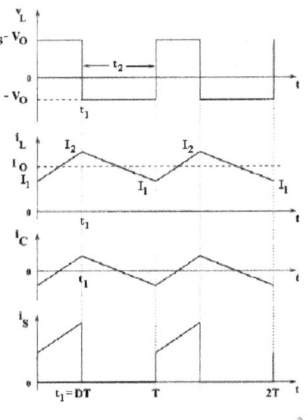

$$\therefore t_1 = L\frac{\Delta I}{V_s - V_o} \quad \longrightarrow \quad (1)$$

Mode (2)

- When S is off at $t=t_1$

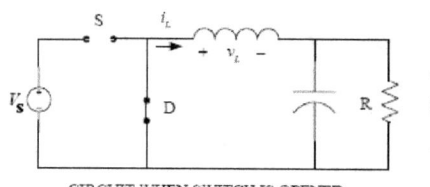

CIRCUIT WHEN SWITCH IS OPENED

- D conducts and the inductor current (i_L) continues to flow through L, C, R and D.
- Assuming i_L falls from I_2 to I_1 in time t_2.

$$-V_o = L\frac{di}{dt} = L\frac{I_1 - I_2}{t_2} = -L\frac{I_2 - I_1}{t_2}$$

$$-V_o = -L\frac{\Delta I}{t_2} \qquad \therefore t_2 = L\frac{\Delta I}{V_o} \quad \longrightarrow \quad (2)$$

$$\therefore \text{Ripple Current}, \Delta I = \frac{t_1}{L}(V_s - V_o) = \frac{t_2}{L}V_o$$

$$t_1 V_s - t_1 V_o = t_2 V_o$$

$$t_1 V_s = V_o (t_1 + t_2)$$

$$t_1 V_s = V_o T$$

$$V_o = V_s \frac{t_1}{T} = DV_s \longrightarrow (3)$$

- Assuming a lossless circuit ($P_i = P_o$)

$$\because P_i = P_o$$
$$\therefore V_s I_s = V_o I_o = D V_s I_o$$

$\therefore$ Average input current $(I_s) = D I_o$ $\longrightarrow$ (4)

$$\because T = t_1 + t_2 = L\frac{\Delta I}{V_s - V_o} + L\frac{\Delta I}{V_o}$$

$$T = \frac{L \Delta I (V_o + V_s - V_o)}{V_o (V_s - V_o)} = \frac{L \Delta I V_s}{V_o (V_s - V_o)}$$

$$\therefore \Delta I = \frac{V_o (V_s - V_o) T}{L V_s} = \frac{V_o (V_s - V_o)}{f L V_s}$$

$$\because V_o = DV_s$$

$$\therefore \Delta I = \frac{DV_s(V_s - DV_s)}{fLV_s}$$

$$\Delta I = \frac{D(1-D)V_s}{fL} \longrightarrow (5)$$

- The Capacitor ripple voltage $(\Delta V_c) = \dfrac{\Delta I}{8fC} \longrightarrow (6)$

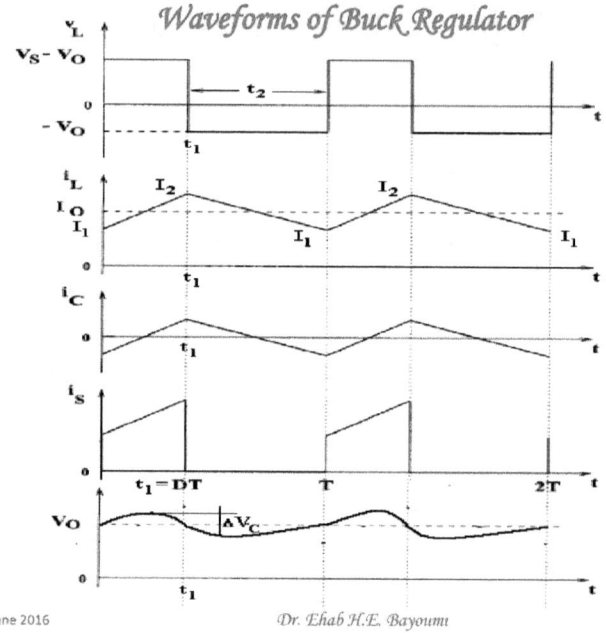

EEL 2003   Electrical / Electronics Department   LO4

Converter circuit topologies

A large number of dc-dc converter circuits are known that can increase or decrease the magnitude of the dc voltage and/or invert its polarity [1-5]. Figure 4 illustrates several commonly used dc-dc converter circuits, along with their respective conversion ratios. In each example, the switch is realized using a power MOSFET and diode; however, other semiconductor switches such as IGBTs, BJTs, or thyristors can be substituted if desired.

The first converter is the buck converter, which reduces the dc voltage and has conversion ratio $M(D) = V_o / V_s = D$.

In a similar topology known as the boost converter, the positions of the switch and inductor are interchanged. This converter produces an output voltage V that is greater in magnitude than the input voltage V_g. Its conversion ratio is $M(D) = V_o / V_s = 1/(1 - D)$.

In the buck-boost converter, the switch alternately connects the inductor across the power input and output voltages. This converter inverts the polarity of the voltage, and can either increase or decrease the voltage magnitude. The conversion ratio is $M(D) = V_o / V_s = - D/(1 - D)$.

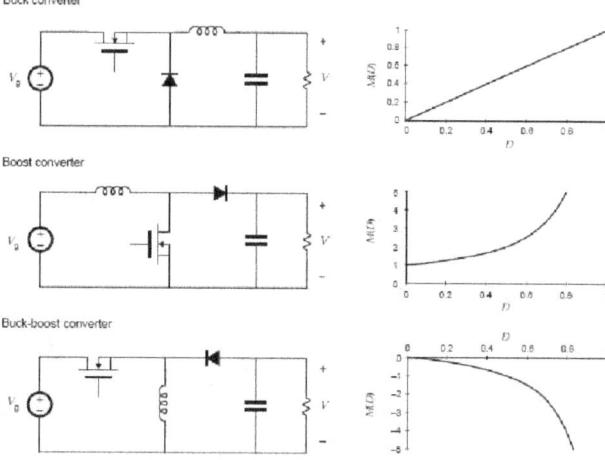

Figure 3 illustrates one way to realize the switch network in the buck converter, using a power MOSFET and diode. A gate drive circuit switches the MOSFET between the conducting (on) and blocking (off) states, as commanded by a logic signal δ(t). When δ(t) is high (for $0 < t < DT_s$), then MOSFET $Q1$ conducts with negligible drain-to-source voltage. Hence, $v_s(t)$ is approximately equal to V_g, and the diode is reverse-biased. The positive inductor current $I_L(t)$ flows through the MOSFET. At time $t = DT_s$, δ(t) becomes low, commanding MOSFET $Q1$ to turn off. The inductor current must continue to flow; hence, $I_L(t)$ forward-biases diode $D1$,
and $v_s(t)$ is now approximately equal to zero. Provided that the inductor current $I_L(t)$ remains positive, then diode $D1$ conducts for the remainder of the switching period. Diodes that operate in the manner are called *freewheeling diodes*.

Since the converter output voltage $v(t)$ is a function of the switch duty cycle D, a control system can be constructed that varies the duty cycle to cause the output voltage to follow a given reference v_r. Figure 3 illustrates the block diagram of a simple converter feedback system. The output voltage is sensed using a voltage divider, and is compared with an accurate dc reference voltage v_r. The resulting error signal is passed through an op-amp compensation network. The analog voltage $v_c(t)$ is next fed into a *pulse-width modulator*. The modulator produces a switched voltage waveform that controls the gate of the power MOSFET $Q1$. The duty cycle D of this waveform is proportional to the control voltage $v_c(t)$. If this control system is well designed, then the duty cycle is automatically adjusted such that the converter output voltage v follows the reference voltage v_r, and is essentially independent of variations in v_g or load current.

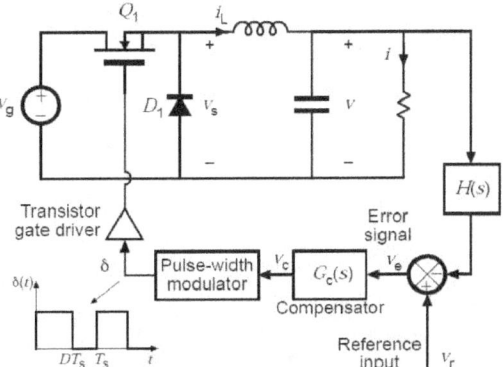

Figure 3. Realization of the ideal SPDT switch using a transistor and freewheeling diode. In addition, a feedback loop is added for regulation of the output voltage.

2- Boost Regulator

➢ The boost regulator is a step up chopper, has an average voltage greater than the input voltage.

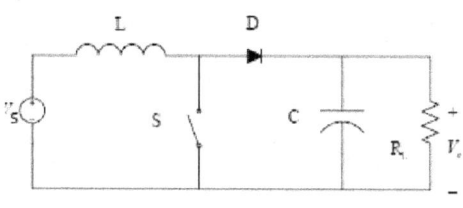

CIRCUIT OF BOOST CONVERTER

▪ There are 2-modes of operation, mode (1) when the transistor (switch S) is ON or closed, and mode (2) when the transistor (switch S) is OFF or open.

Mode (1)

❏ When S is on at $t=0$

▪ Input current flows through filter inductor (L), and the switch (S).

$$e_L = L \frac{di}{dt}$$

➢ Assuming i_L rises from I_1 to I_2 in time t_1.

$$e_L = L \frac{(I_2 - I_1)}{(t_1 - 0)} = L \frac{I_2 - I_1}{t_1}$$

$$\therefore e_L = V_s = L \frac{I_2 - I_1}{t_1} = L \frac{\Delta I}{t_1}$$

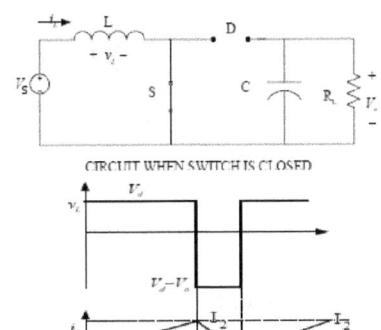

CIRCUIT WHEN SWITCH IS CLOSED

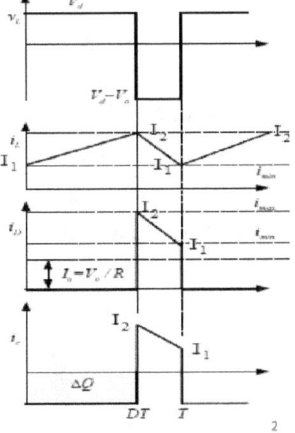

$$\therefore t_1 = L\frac{\Delta I}{V_s} \quad\longrightarrow\quad (1)$$

Mode (2)

➢ When S is off at $t=t_1$

CIRCUIT WHEN SWITCH IS OPENED

➢ The input current (i_s) flows through L, D, C and the load.
➢ The inductor current(i_L) decreases till the switch (S) is **on** again.
➢ The energy stored in L is transferred to the load.
➢ Assuming i_L falls from I_2 to I_1 in time t_2.

$$V_s = V_o \pm e_L$$

$$V_s - V_o = e_L = L\frac{(I_1 - I_2)}{t_2} = -L\frac{(I_2 - I_1)}{t_2}$$

$$V_s - V_0 = -L\frac{\Delta I}{t_2} \qquad \therefore t_2 = L\frac{\Delta I}{V_o - V_s} \longrightarrow (2)$$

$$-(V_o - V_s) = -L\frac{\Delta I}{t_2}$$

$$\therefore \text{Ripple Current}, \Delta I = \frac{t_1}{L}V_s = \frac{t_2}{L}(V_o - V_s)$$

$$t_1 V_s = t_2 V_o - t_2 V_s$$

$$(t_1 + t_2)V_s = V_o t_2$$

$$V_o = V_s \frac{(t_1 + t_2)}{t_2} = V_s \frac{T}{T - t_1} = V_s \frac{1}{1 - t_1/T} = V_s \frac{1}{1 - D}$$

$$V_o = \frac{V_s}{1 - D} \longrightarrow (3)$$

> Assuming a lossless circuit ($P_i = P_o$)

$$\because P_i = P_o$$

$$\therefore V_s I_s = V_o I_o = \frac{V_s}{1-D} I_o$$

$$\therefore \text{Average input current}(I_s) = \frac{I_o}{1-D} \quad \longrightarrow \quad (4)$$

$$\because T = t_1 + t_2 = L\frac{\Delta I}{V_s} + L\frac{\Delta I}{V_o - V_s}$$

$$T = \frac{L \Delta I (V_o - V_s + V_s)}{V_s (V_o - V_s)} = \frac{L \Delta I V_o}{V_s (V_o - V_s)}$$

$$\therefore \Delta I = \frac{V_s (V_o - V_s) T}{L V_o} = \frac{V_s (V_o - V_s)}{f L V_o} \quad \longrightarrow \quad (5)$$

$$\therefore \frac{V_s}{V_o} = (1-D) \qquad \longrightarrow \quad (6)$$

- Substitute by Eq. (6) in Eq. (1):

$$\therefore \Delta I = \frac{1}{fL} V_s \left(\frac{V_o - V_s}{V_o} \right) = \frac{1}{fL} V_s (1-(1-D))$$

$$\Delta I = \frac{DV_s}{fL} \qquad \longrightarrow \quad (7)$$

- During Mode (1), the capacitor current (i_c) supplies the load,

$$\therefore i_c = I_o \qquad for \ 0 < t < t_1$$

$$\Delta V_c = \frac{1}{C}\int_0^{t_1} I_o \, dt = \frac{I_o t_1}{C} = \frac{I_o}{C} L \frac{\Delta I}{V_s}$$

$$\Delta V_c = \frac{I_o}{C} \frac{L}{V_s} \frac{V_s D}{fL}$$

- The Capacitor ripple voltage $(\Delta V_c) = \dfrac{I_o D}{fC}$ ⟶ (8)

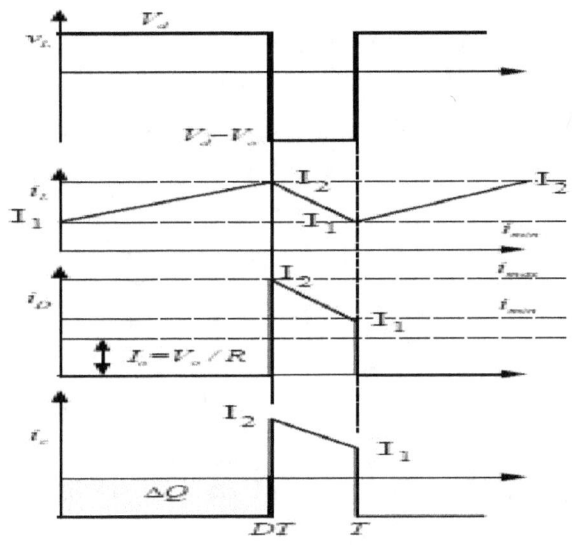

3- Buck-Boost Regulator

➤ The buck-boost regulator is may be a step up or step down chopper, has an average voltage greater or less than the input voltage.

▪ The output voltage polarity is opposite to that of the input voltage.

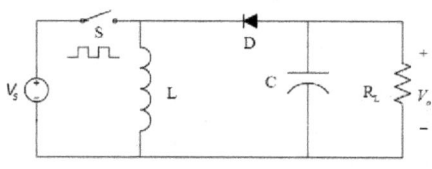

CIRCUIT OF BUCK-BOOST CONVERTER

▪ There are 2-modes of operation, mode (1) when the transistor (switch S) is ON or closed, and mode (2) when the transistor (switch S) is OFF or open.

Mode (1)

- When S is on at $t=0$

- Input current flows through filter inductor (L), and the switch (S).

$$e_L = L \frac{di}{dt}$$

➤ Assuming i_L rises from I_1 to I_2 in time t_1.

$$e_L = L \frac{(I_2 - I_1)}{(t_1 - 0)} = L \frac{I_2 - I_1}{t_1}$$

$$\therefore e_L = V_s = L \frac{I_2 - I_1}{t_1} = L \frac{\Delta I}{t_1}$$

CIRCUIT WHEN SWITCH IS CLOSED

$$\therefore t_1 = L\frac{\Delta I}{V_s} \quad\longrightarrow\quad (1)$$

Mode (2)

➢ *When S is off at $t=t_1$*

CIRCUIT WHEN SWITCH IS OPENED

➢ The input current (i_s) = 0, in this mode of operation.
➢ The inductor current(i_L) decreases till the switch (S) is **on** again.
➢ The energy stored in L is transferred to the load.
➢ Assuming i_L falls from I_2 to I_1 in time t_2.

$$e_L = -V_o$$
$$e_L + V_c = e_L - V_o = 0$$
$$V_o = e_L = L\frac{(I_1 - I_2)}{t_2} = -L\frac{(I_2 - I_1)}{t_2}$$

$$\therefore t_2 = -L\frac{\Delta I}{V_o} \quad \longrightarrow \quad (2)$$

$$\therefore \text{Ripple Current}, \Delta I = \frac{t_1}{L}V_s = -\frac{t_2}{L}V_o$$

$$V_o = -\frac{t_1}{t_2}V_s = -\frac{Lt_1}{T-t_1}V_s$$

$$V_o = -\frac{\frac{t_1}{T}}{1-\frac{t_1}{T}}V_s = \frac{-D}{1-D}V_s$$

$$V_o = \frac{D}{D-1}V_s \quad \longrightarrow \quad (3)$$

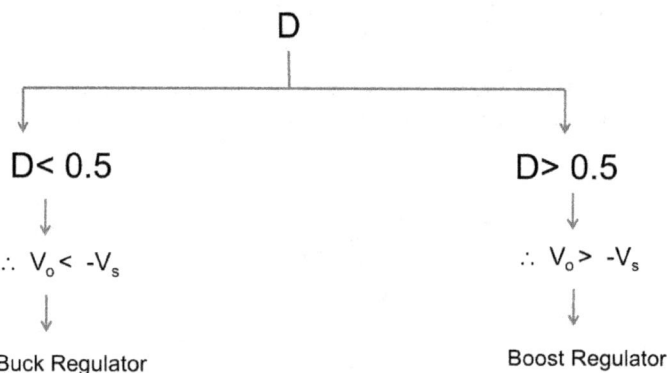

- Assuming a lossless circuit ($P_i = P_o$)

$$\because P_i = P_o$$

$$\therefore V_s I_s = -V_o I_o = \frac{D}{1-D} V_s I_o$$

$$\therefore \text{Average input current}(I_s) = \frac{D}{1-D} I_o \quad \longrightarrow \quad (4)$$

$$\because T = t_1 + t_2 = L\frac{\Delta I}{V_s} - L\frac{\Delta I}{V_o}$$

$$T = \frac{L \Delta I (V_o - V_s)}{V_s V_o}$$

$$\therefore \Delta I = \frac{V_s V_o T}{L(V_o - V_s)} = \frac{V_s V_o}{fL(V_o - V_s)} \quad \longrightarrow \quad (5)$$

$$\therefore V_o = \frac{D}{D-1} V_s \qquad (6)$$

➢ Substitute by Eq. (6) in Eq. (5):

$$\therefore \Delta I = \frac{V_s(\frac{D}{(D-1)} V_s)}{fL(\frac{D}{(D-1)} V_s - V_s)} = \frac{(\frac{D}{D-1}) V_s}{fL(\frac{D}{D-1} - 1)} = \frac{(\frac{D}{D-1}) V_s}{fL(\frac{D-D+1}{D-1})}$$

$$\Delta I = \frac{D V_s}{fL} \qquad (7)$$

➢ During Mode (1), the capacitor current (i_c) supplies the load,

$$\therefore i_c = I_o \qquad for \ 0 < t < t_1$$

$$\Delta V_c = \frac{1}{C}\int_0^{t_1} I_o \, dt = \frac{I_o t_1}{C} = \frac{I_o}{C}\frac{L}{V_s} \; \Delta I$$

$$\Delta V_c = \frac{I_o}{C}\frac{L}{V_s}\frac{V_s D}{fL}$$

- The Capacitor ripple voltage $(\Delta V_c) = \dfrac{I_o D}{fC}$ ⟶ (8)

Waveforms of Buck-Boost Regulator

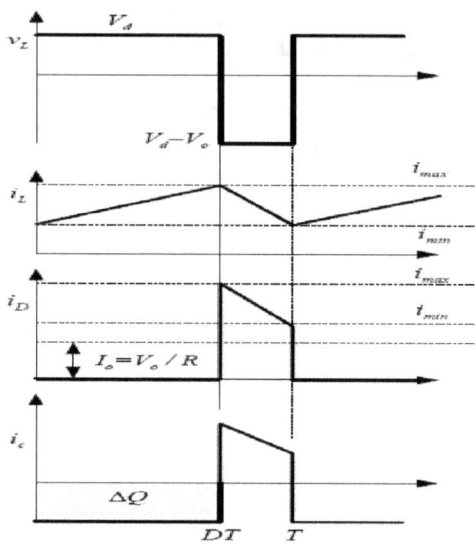

Examples

1- In a buck-boost regulator the input voltage 12 V, a duty cycle of 0.6, an average load current 1.5 A, and switching frequency 25 kHz. If the filter inductance and capacitance are 250 µH and 220 µF respectively, determine:
 1. the average output voltage,
 2. peak-to-peak output ripple voltage, and
 3. peak-to-peak ripple of the inductor current.

www.ingramcontent.com/pod-product-compliance
Lightning Source LLC
Chambersburg PA
CBHW052302220526
45471CB00001B/460